Micromechanisms in particle-hardened alloys

Cambridge Solid State Science Series

EDITORS:

Professor R. W. Cahn
Applied Sciences Laboratory, University of Sussex

Professor M. W. Thompson
School of Mathematical and Physical Sciences, University of Sussex

Professor I. M. Ward
Department of Physics, University of Leeds

J. W. MARTIN

Lecturer in Metallurgy and Science of Materials, and Fellow of St. Catherine's College, Oxford

Micromechanisms in particle-hardened alloys

CAMBRIDGE UNIVERSITY PRESS

CAMBRIDGE

LONDON   NEW YORK   NEW ROCHELLE

MELBOURNE   SYDNEY

CAMBRIDGE UNIVERSITY PRESS
Cambridge, New York, Melbourne, Madrid, Cape Town,
Singapore, São Paulo, Delhi, Tokyo, Mexico City

Cambridge University Press
The Edinburgh Building, Cambridge CB2 8RU, UK

Published in the United States of America by
Cambridge University Press, New York

www.cambridge.org
Information on this title: www.cambridge.org/9780521295802

First published 1980
Re-issued 2013

A catalogue record for this publication is available from the British Library

Library of Congress cataloguing in publication data
Martin, John Wilson
Micromechanisms in particle-hardened alloys
(Cambridge solid state science series)
Bibliography: p. 193
Includes index
1. Alloys.   2. Precipitation hardening
3. Physical metallurgy.   I. Title
TN693.I7M328   669′.95   78–74011

ISBN 978-0-521-22623-3 Hardback
ISBN 978-0-521-29580-2 Paperback

Contents

Preface

A great number of metallic materials in engineering use owe their strength to the presence in their microstructure of particles of a hard, finely-divided phase. Theoretical models of strengthening mechanisms in this type of system have been proposed and developed over the past thirty years or more, and most modern undergraduate textbooks of physical metallurgy introduce the student to some of these ideas. It is only comparatively recently, however, that attempts have been made to interpret the behaviour of 'real' materials, rather than of model systems, in terms of these theories, and it is hoped that the present text will give an indication of the degree of success of these attempts.

An elementary knowledge of physical metallurgy by the reader has been assumed. The book develops the theories relating the deformation and fracture behaviour of particle-hardened alloys to their microstructure, and also attempts, where possible, to account for experimental observations on a number of practical commercial alloys (as well as model materials) in the light of such theories. This approach thus emphasizes the importance of scientific, rather than empirical, methods when attempting to develop improved materials of this type. The level of the text is intended to be appropriate for final-year undergraduate reading, but it is hoped that by the inclusion of much fairly recent work, the book will prove of value to many research workers in the field of alloy development.

The opening chapter discusses how the microstructures of these alloys are formed and how precipitate distribution may be controlled. The reader is also introduced to the quantitative metallography of microstructures containing a dispersed second phase. Chapter 2 considers yield and work-hardening in the absence of recovery. Of all the mechanical properties considered in the book, it is certainly with the yield stress that the greatest success has been achieved in the correlation of theory with observation. It was not felt appropriate to develop from first principles the more sophisticated theories of work-hardening, although an outline is given of their structure. Full literature references are provided for the more advanced reader. Chapter 3 is concerned with micromechanisms of fracture in two-phase materials, including a consideration of fracture toughness and of fatigue behaviour. The book concludes (chapter 4) with an account of mechanisms at elevated temperatures. This final chapter considers not only recrystallization mechanisms, but also mechanisms of yielding, work-hardening and creep. Again, in dealing with these latter topics, it is hoped that the more advanced reader will be led to the extensive background literature by the selection of references included.

The author would like to acknowledge the important contributions made by colleagues in Oxford and elsewhere in helping him to formulate the ideas set out in the following pages. He is particularly grateful for the stimulus provided by the long series of graduate and undergraduate students with whom it has been his pleasure and privilege to work. He is also indebted to the scientists who have given permission for the use of their micrographs, and to the publishers of scientific journals for permission to reproduce figures.

Oxford, March 1979   J. W. Martin

1 The structure of particle-hardened alloys

1.1 Dispersed phases present in metals

In most theoretical discussions of strengthening mechanisms in particle-hardened alloys, attention is usually confined to the interaction of glide dislocations with finely-dispersed precipitates typically 10 nm in size. In real alloys of this type, however, dispersions of coarser particles also exist, which may play an important role in the deformation behaviour and particularly in the fracture behaviour of the alloy. The types of dispersed phases that may be present can be conveniently classified into three families.

1.1.1 Hardening precipitates

These may range in size from, say, 1 to 100 nm. Chapters 1 and 2 are particularly concerned with the formation of such dispersions and their effect upon the yield and work-hardening behaviour of the alloy. In steels and in age-hardening non-ferrous alloys the particles are formed by precipitation from supersaturated solid solution, and this constitutes by far the most commonly used technique in use for producing a dispersed second phase. Other methods include diffusion reaction techniques, such as nitriding of steel or internal oxidation of, for example, copper alloys, and powder-metallurgical techniques.

Internal oxidation has been used fairly extensively in academic studies to produce single crystals containing fine dispersions of oxide particles, which form admirable 'model' systems for basic study. An alloy suitable for this treatment consists of a dilute solid solution of a base metal in a more noble metal. When the alloy is heated under oxidizing conditions, oxygen diffuses into the alloy, producing a dispersion of the oxide of the base metal in a matrix of the noble metal. Typical alloys appropriate for this treatment would be copper containing, say, 0.1wt% of silicon, aluminium or beryllium which form dispersions of SiO_2, Al_2O_3 and BeO respectively. These phases will have high thermodynamic stability in comparison with phases produced by precipitation from supersaturated solid solution, yet are more finely divided than is normally achieved by employing powder-metallurgical techniques of physically mixing powdered metals and oxides to produce composites.

Physical mixing of metal/oxide composites followed by compaction and sintering does not in general lead to the formation of structures consisting of uniform dispersions of finely-divided oxide particles. Even if the oxides are evenly distributed (a state extremely difficult, if not impossible, to achieve by mixing methods) the minimum interparticle spacing could never be less than that of the metal powder granule size itself. Since the enhancement of properties by dispersions increases with decrease of the interparticle spacing, this size effect is a severe limitation to the utility of 'straight' powder-metallurgical methods in producing dispersion-hardened metal refractory systems.

One of the successful powder-metallurgical products in this field is the material known as SAP (sintered aluminium powder). This is made by the hot compaction of flake-shaped aluminium powder (of flake thicknesses in the range 0.01 to 0.1 μm) which contains up to about 15wt% alumina formed during the manufacture of the powder. The sintered products consist of a dispersion of alumina in aluminium, and they possess exceptional strength and thermal stability. The success of SAP arises from the extremely small aluminium-flake thickness, which provides the required small oxide–oxide distance in the product. Nickel-based alloys containing dispersed oxides such as ThO_2 (TD-Nickel and TD-Nichrome) have been successfully produced by precipitating a hydrous, oxygen-containing compound of the matrix metal on to dispersoid particles in the form of a colloidal aquasol. This is followed by reduction, compaction, sintering and densification which produces structures containing good dispersions of fine oxide particles.

1.1.2 Coarse residual particles

When considering such particles, which are larger than 1 μm in size, it is convenient to discuss separately the occurrence of these coarse inclusions in *steels* and those encountered in *non-ferrous alloys*.

Inclusions in steels

The origin and constitution of non-metallic inclusions in steels have been the subject of intense study over many years, as it has long been recognized that they are a potential source of weakness. It is for this reason that strenuous efforts are now made in the production of 'clean' steels for many applications. The three main sources of non-metallic inclusions are:

(i) deoxidation, and the segregation of the products of deoxidation;

(ii) the presence of sulphur and phosphorous, and the segregation of their compounds;

(iii) extraneous sources, including the trapping of slag and eroded refractory materials within the molten steel.

Considerable attention has been paid to the deformability of different types of non-metallic inclusions during hot-working. Silicate inclusions are known to deform extensively above about 1000 °C, but below this temperature they either deform little or shatter and disperse, as do the higher-melting-point oxide inclusions. The other major inclusion-type is manganese sulphide, which is always present to some degree to 'tie up' the residual sulphur present in the steel which would otherwise be present as a low-melting-point FeS eutectic phase. The deformability of MnS increases relative to that of steel with decreasing rolling temperature of the steel. Very elongated stringers of MnS are encountered in wrought steels, and together with the silicate inclusions are usually considered to be the major cause of poor ductility.

Non-ferrous alloys

Coarse insoluble particles are formed during casting of non-ferrous alloys, and although these may be broken up and distributed more uniformly through the structure by hot-working of the ingot, they are again recognized as a potential source of weakness in the material. Commercial *aluminium alloys* contain from about 1% to 5% by volume of large iron- or silicon-rich inclusions, and may also contain copper-bearing particles arising from non-equilibrium microsegregation during solidification. Iron is, of course, the principal impurity in bauxite, so that its presence is not unexpected in the final product.

1.1.3   Intermediate-sized dispersoids

These may range in size from, say 0.1 to 1 μm, and we will again take as our example aluminium alloys (in which such particles are commonly found), although they may occur in many materials. Chromium, zirconium, or manganese is added to many commercial wrought aluminium alloys. The element usually remains in solution during casting, but during fabrication these alloys are normally given a so-called 'homogenization' heat-treatment at relatively high temperature. The heat-treatment results in the formation of precipitates of intermetallic compounds containing chromium, zirconium or manganese, whose size and spacing depend upon the temperature and time of homogenization. An example of such a dispersion in an Al-Mg-Si alloy is shown in fig. 1.1.

These dispersions have pronounced retardation effects upon the response of the alloy to recrystallization and grain growth, and are also known in some alloy systems to reduce the tendency for intergranular embrittlement in the fully-aged condition. We will return later to a consideration of their effect upon deformation and fracture processes in age-hardenable alloys.

It is clear, therefore, that in industrial alloys of practical significance, several families of precipitates of differing ranges of particle size are likely to be present and to have an effect upon the deformation and/or fracture

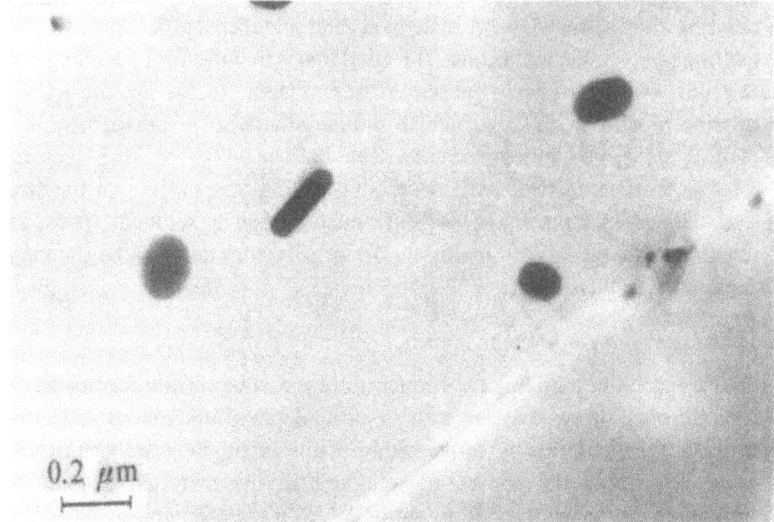

Fig. 1.1. Electron micrograph of an Al–0.6wt% Mg–1wt% Si–0.6wt% Mn alloy aged to peak hardness. The Mg_2Si phase is present as noncoherent particles at the grain boundaries and as the fine, coherent hardening phase within the grains. Note the narrow precipitate-free zone adjacent to the grain boundary. The Mn is present as dispersoids of the α-$Al_{12}Mn_3Si$ phase, typically 0.1 μm in size. (Courtesy of K. C. Prince.)

behaviour of the material. We will next consider the kinetics of the nucleation and growth of precipitates in solids, how such precipitates are distributed in the microstructure, and examine ways in which this distribution may be controlled.

This chapter will conclude with a consideration of how the distribution of precipitates in alloys may be quantitatively defined.

1.2   Precipitation in solids

Christian (1975) has reviewed the subject of precipitation in solids comprehensively, and this topic is dealt with at an introductory level in standard undergraduate texts on phase transformations in condensed systems. We will adopt a generalized mechanistic approach to account for the microstructures observed in metallic systems containing precipitates.

The decomposition of a phase into one or more phases may be divided into three stages: (i) the formation of nuclei of the new phase; (ii) the growth of these nuclei; and (iii) the coarsening of the precipitate without change in its volume fraction. Stage (i) may itself occur in two ways: if small concentration fluctuations may occur spontaneously, the reaction may proceed by *spinodal decomposition*; if all small fluctuations tend to decay, there is said to be a nucleation barrier, and we will consider first the energy relations which contribute to the magnitude of this barrier, and then the kinetic factors which determine the rate at which the barrier is overcome and successful nucleation occurs.

1.2.1   Nucleation from supersaturated solid solution

Energy relations

We will initially consider a situation involving the homogeneous nucleation of a second phase, i.e. nucleation which occurs without the benefit of pre-existing heterogeneities in the system. The basic ideas of nucleation theory were originally expressed by Gibbs (1878), namely that work is necessary for the formation of the *surface* of a new phase.

Let us assume that, in a metastable α-phase, a β-region forms consisting of n atoms. If σ is the specific surface energy of the $\beta\alpha$ interface and Δg_c the specific chemical free energy of the β-phase, the balance of surface energy $(a\sigma n^{\frac{2}{3}})$ and the chemical energy due to the transformation in the new structure $(\Delta g_c n V)$ is

$$\Delta G = a\sigma n^{\frac{2}{3}} + \Delta g_c n V,$$

where V is the atomic volume, and a depends upon the shape of the β-region, which in the simplest case is a sphere. Transformations in the solid state are also usually associated with a change in specific volume, which leads to distortions during nucleation. An elastic strain-energy term (g_e) should thus be included in the energy balance to give

$$\Delta G = a\sigma n^{\frac{2}{3}} + (\Delta g_c + g_e) n V. \tag{1.1}$$

The variation in ΔG with n will therefore be of the form shown in fig. 1.2 and the condition for continued growth of an embyro is that the

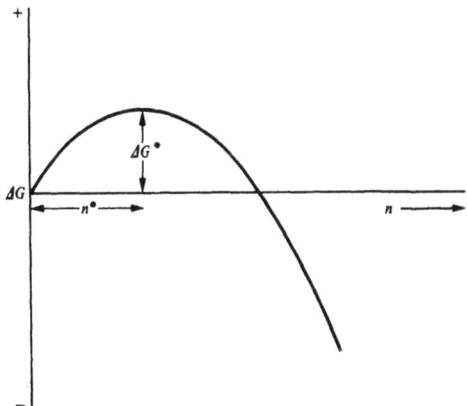

Fig. 1.2. Free energy (ΔG) of a precipitate as a function of the number of atoms it contains (n).

number of atoms it contains should exceed n^*, where $d\Delta G/dn = 0$, i.e.

$$n^* = \left(\frac{2a\sigma}{-3V(\Delta g_c + g_e)}\right)^3. \tag{1.2}$$

The critical free energy, or activation energy for nucleation, is given by

$$\Delta G^* = \tfrac{1}{3}\sigma n^{*\frac{2}{3}}. \tag{1.3}$$

For a spherical nucleus of radius r, the surface area is given by $4\pi r^2$ $= an^{\frac{2}{3}}$. Since $n = 4\pi r^3/3V$, we may substitute in equations (1.2) and (1.3) to obtain

$$r^* = \frac{2\sigma}{-(\Delta g_c + g_e)} \tag{1.4}$$

and

$$\Delta G^* = \frac{16\pi\sigma^3}{3(\Delta g_c + g_e)^2}. \tag{1.5}$$

Below the equilibrium temperature Δg_c becomes negative and increases roughly linearly with increasing undercooling (ΔT). The values of σ and g_e are assumed to be independent of temperature.

The temperature dependence of r^* and ΔG^* may be assessed in terms of the temperature dependence of Δg_c, so we can write as a first approximation

$$r^* \propto 1/\Delta T, \quad \Delta G^* \propto 1/\Delta T^2. \tag{1.6}$$

Basic kinetic theory

A steady-state nucleation rate, $\dot{N}_V$, may be defined as the number of stable nuclei produced in unit time in unit volume of untransformed solid. The theory assumes that the atomic fluctuations which give rise to the embryos are present in statistical equilibrium, so that if the number of atoms, n, in these fluctuations is much smaller than the total number of atoms, n_0, it follows from (1.1) that

$$n = n_0 \exp(-\Delta G/kT),$$

where k is the Boltzmann constant and T the absolute temperature. An embryo which contains a larger number of atoms than n^* (fig. 1.2) can grow with decreasing free energy and is called a nucleus, and it follows that the nucleation rate $\dot{N}_V$ is proportional to $\exp(-\Delta G^*/kT)$. The rate at which individual nuclei grow will also be dependent on the frequency with which atoms adjacent to the nucleus can join it, and this will be proportional to the volume diffusivity, D. One may therefore write a simplified representation of the rate of nucleation of a precipitate

$$\dot{N}_V = KD \exp \frac{-A\sigma^3/(\Delta g_c + g_e)^2}{kT}, \tag{1.7}$$

where A is a geometrical constant and K another constant.
This relationship accounts for the observed minima in the incubation time in temperature-time-transformation (TTT) curves. Because of the high value of the energy of activation for diffusion (1 to 4 eV for substitutional atoms) the value of D, and thus the rate of nucleation becomes very low at low temperatures. At temperatures close to equilibrium, $\dot{N}_V$ becomes low because $\Delta G^* \to \infty$ (since $\Delta g_c \to 0$).

The precipitation of face-centred cubic (fcc) cobalt from fcc copper-cobalt alloys is one of the few systems which decompose by homogeneous nucleation. Servi & Turnbull (1966) have studied this system, and found that the theory predicts the temperature (at fixed composition, or vice-versa) for a given nucleation rate very accurately. In (1.7), g_e is neglected and Δg_c is taken as the product of kT and the natural log of the supersaturation, i.e. $kT \ln c/c_e$, where c is the cobalt content of a particular (supersaturated) alloy and c_e the equilibrium solubility of cobalt at the precipitation temperature. Servi & Turnbull presented their data in an activation plot. This plot is reproduced as fig. 1.3. The effective nucleation time is denoted by $\bar{t}$, so N_p (number of particles) divided by $\bar{t}$ is nucleation rate, and the plot shown in fig. 1.3 is arrived at by rearranging (1.7) and substituting for Δg_c. If σ is independent of temperature the plot should give a straight line, and Servi & Turnbull drew a straight line through their data

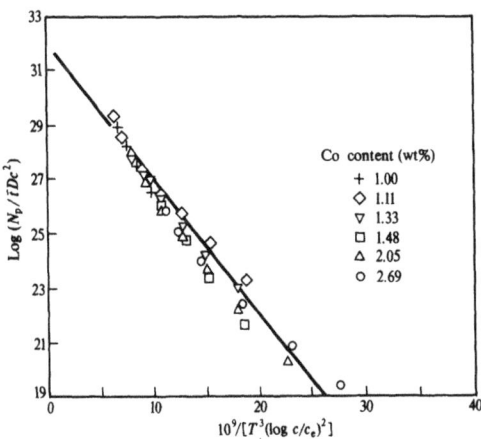

Fig. 1.3. Activation plot of data relating to the coherent nucleation of Co from a dilute solution of Co in Cu. (After Servi & Turnbull, 1966.)

and compared the slope and intercept to predictions of nucleation theory. The slope of fig. 1.3 corresponds to a value of σ of 0.23 J m^{-2}, which agrees well with a calculated value of 0.20 J m^{-2}, but the intercept on the vertical axis (corresponding to the pre-exponential term in (1.7)) of 32 is less than the anticipated value of 36, which may be due to experimental uncertainties and/or a curvature in the plot due to ignored surface entropy effects.

The behaviour of the Cu–Co system is not typical of most precipitated alloys of practical interest. Instead of the homogeneous nucleation of an equilibrium phase, metastable phases are commonly observed, and furthermore nucleation sites are very commonly associated with lattice defects in the matrix - in other words, one usually encounters *heterogeneous nucleation*.

1.2.2   Heterogeneous nucleation

In supersaturated solid solutions the following types of lattice defects can be expected;

0-dimensional faults - vacancies and interstitialcies;

1-dimensional faults - dislocations;

2-dimensional faults - grain and twin boundaries, stacking faults, antiphase domain boundaries, and possibly interphase boundaries.

The rate of heterogeneous nucleation is again proportional to exp $(-\Delta G^*/kT)$ and the magnitude of ΔG^* is lower for the heterogeneous nucleation than for homogeneous nucleation due to the reduction in one or both of σ and g_e (1.1) caused by the interaction of the defects and the critical nucleus.

The crystal structures of the matrix and the precipitate must also be considered, because quite different structures and hence energies of the interface will arise from this factor. There are three important cases to define.

(i) *Coherent nucleation*: if the crystal structures and lattice parameters of both phases are closely similar, this will occur (fig. 1.4a). The precipitation of the Ni$_3$Al-phase in Ni is of this type, the dispersed particles being in parallel crystallographic orientation with the matrix.

(ii) *Semicoherent nucleation* arises if the α- and β-phases are related in such a way that an interface can be built from well-defined defects, such as interfacial dislocations (fig. 1.4b). The θ'-phase in the Al–Cu system is a familiar example of this type.

(iii) *Noncoherent nucleation* arises if the structures of the α- and β-phases are so different that the interface has a structure similar to that of a high-angle grain boundary (fig. 1.4c). The θ-phase in the Al–Cu system is of this type.

There is a limited number of fixed orientations between the α- and

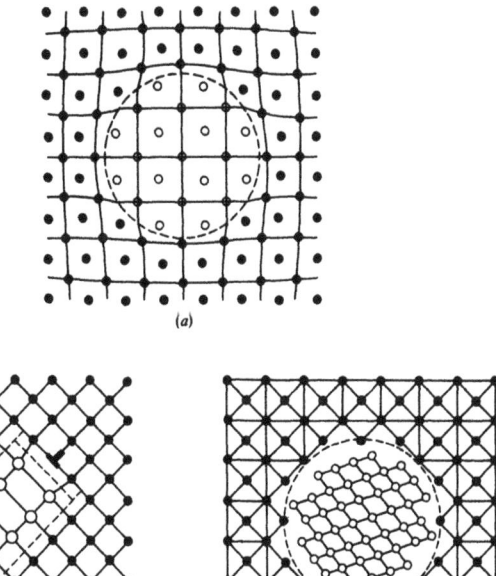

Fig. 1.4. (*a*) Fully coherent precipitate showing coherency strains in matrix. (*b*) Partially coherent precipitate showing interface dislocations. (*c*) Non-coherent precipitate.

β-phases which will allow coherent or semicoherent nucleation. In non-coherent nucleation, however, a random orientation between the two phases can be expected.

Heterogeneous nucleation on point defects

Quenched-in vacancies can affect the rate of precipitate nucleation in two ways. Firstly, they can increase the diffusion rate of solute atoms and thereby increase the growth rate of precipitate embryos. Secondly, vacancies may become an integral part of the nucleus and thereby reduce the nucleation barrier to precipitate formation. It is this latter function we will consider further for the moment. The nucleation of a misfitting precipitate will be facilitated if a number of vacancies accumulate in order to counteract the volume change, thus reducing g_e (see (1.1)). For strain-free nucleation (i.e. $g_e = 0$) the nucleus requires ρ_0 vacancies per atom, ρ_0 is given by

$$\rho_0 = \frac{V^\beta - V^\alpha}{V^\alpha},$$

where V^β is the atomic volume of the precipitate and V^α that of the matrix. Thus, in the presence of vacancies, the strain-energy may be written as $g'_e = g_e (1 - \rho/\rho_0)$, and the energy barrier to nucleation (1.5) now becomes

$$G^* = \frac{16\pi\sigma^3}{3\,[\Delta g_c + \Delta g_v + g_e\,(1 - \rho/\rho_0)]^2},$$

where Δg_v is the change in free energy associated with vacancy precipitation.

There is convincing scientific evidence that excess vacancy concentrations influence precipitation processes. Shepherd (1969) has, for example, shown that a much finer dispersion of carbide precipitates is formed in an irradiated stainless steel specimen than in an equivalent control specimen.

Excess point defects can of course condense to form small dislocation loops, or stacking-fault tetrahedra (depending on the material), and the possibility cannot be discounted that the nucleating sites in irradiated specimens are in fact vacancy aggregates of this type.

Heterogeneous nucleation on dislocations

The first theoretical model for nucleation on dislocations was due to Cahn (1957), who considered a cylindrical nucleus and allowed its misfit to balance the strain-energy of the dislocation so that g_e (1.1) is reduced and $\dot{N}_V$ is increased (1.7). Cahn expresses his result in terms of a dimensionless parameter.

$$\alpha = \frac{\Delta g_c\,Gb^2}{2\pi^2\sigma^2},$$

where G is the shear modulus and b the Burgers vector, and he showed that the effectiveness of dislocations to lower ΔG^* increases with increasing values of α. According to this model, therefore, dislocations with a large Burgers vector are more effective than those with a small one. Experimental investigations have indicated that the mechanism assumed as a basis for this theory is not observed very frequently: it is most realistic in dealing with particles with a spherically symmetrical strain field, such as Co in Cu–Co.

Hornbogen & Roth (1967) made a systematic study of the influence of Δg_e on nucleation in the matrix and on dislocations, by comparing the precipitation kinetics of alloys of comparable supersaturation (i.e. Δg_c = constant) but of varying lattice misfit strain (ϵ) associated with the precipitate. It was always found that the larger the value of ϵ, the larger the time gap between the beginning of nucleation at dislocations and in the matrix (fig. 1.5).

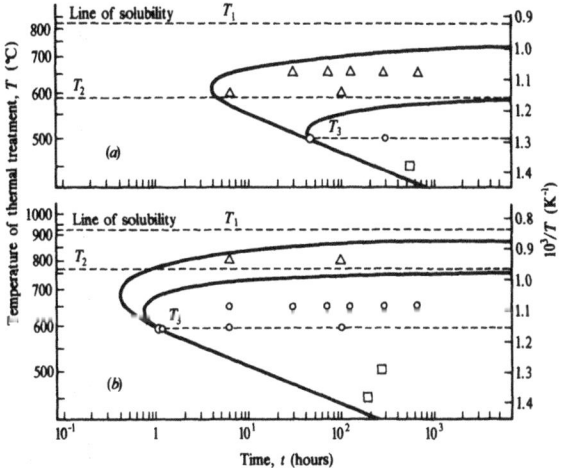

Fig. 1.5. Nucleation diagrams for precipitation in: (a) Ni–6.36wt% Al, misfit strain 0.4%; and (b) Ni–6.32wt% Al, misfit strain 0.34%. T_1 is the equilibrium temperature, T_2 is the temperature above which only dislocation nucleation takes place and T_3 is the temperature below which there is simultaneous nucleation at dislocations and in the matrix. (After Hornbogen & Roth, 1967.)

The greatest effectiveness of a dislocation in lowering the strain-energy and therefore the activation energy of nucleation is to be expected if the nucleus can be formed by dissociation of the dislocation into two partial dislocations. The nucleation event is then involved with a two-dimensional feature (the stacking-fault) rather than the one-dimensional dislocation, and we will discuss this further in § 1.2.3.

Precipitation on dislocations can lead to high precipitate densities if an autocatalytic process operates. For example when carbides of the type $M_{23}C_6$ precipitate in certain austenitic stainless steels the particles are found to lie in strings along $\langle 110 \rangle$ matrix directions. Studies of the early stages of ageing have shown that discrete particles are nucleated on the existing dislocations, and they grow partially coherently with the matrix. In order to accommodate the strains resulting from the difference in atomic volume between the two lattices, prismatic loops of dislocation are 'punched' out by the growing particle into the matrix (fig. 1.6). These punched loops (which have $\langle 110 \rangle$ glide axes) then act as sites for further precipitation of carbide, resulting in the formation of the stringers observed after long ageing times.

Still higher precipitate densities may also be achieved by introducing a high dislocation density into the supersaturated solid solution. These higher densities may arise through transformation strain, as in the case of maraging steels, in which martensite of high dislocation density is formed

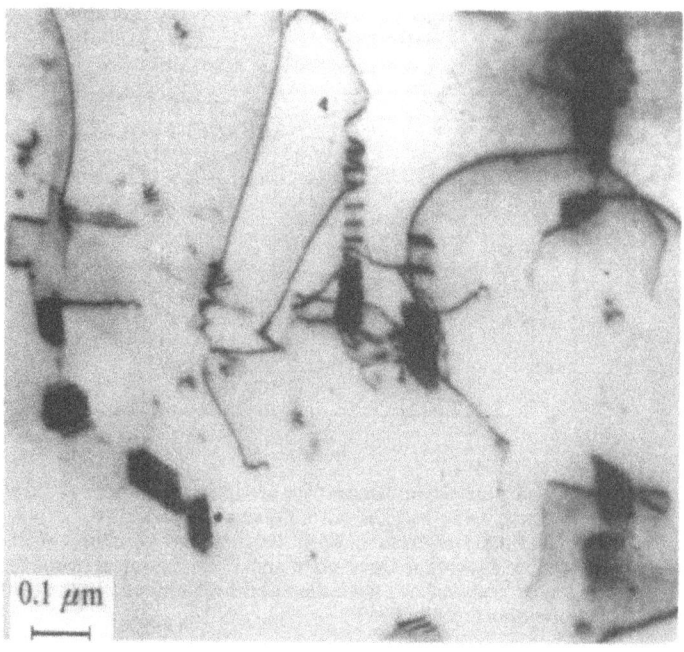

0.1 μm

Fig. 1.6. Electron micrograph of precipitation of iron–chromium carbide $(M_{23}C_6)$ in an austenitic stainless steel after ageing for 5.5 hours at 750 °C. Prismatic dislocation loops are being punched out into the matrix by the growing particles. (Courtesy of M. H. Lewis & B. Hattersley.)

on cooling, or through mechanical working, as in ausforming and other thermomechanical treatment. We will discuss these possibilities in §§ 1.3.3 and 2.4.4.

Heterogeneous nucleation on planar defects

Grain boundaries are effective in increasing $\dot{N}$, the rate of nucleation since, when a precipitate nucleus is formed in a boundary, part of the boundary area is eliminated and this energy is available effectively to reduce the σ-term in (1.7). The energy of a phase boundary is composed of two parts: (i) energy arising from the *structure* of the interface; and (ii) the *chemical* energy due to the difference in composition and order on both sides of the interface. Term (ii) leads to a difference between $\sigma_{\alpha\beta}$ and $\sigma_{\alpha\alpha}$ for an identical structure of grain boundary ($\alpha\alpha$) and phase boundary ($\alpha\beta$). Fig. 1.7 illustrates the form of a β-phase nucleus at an α-grain boundary, which for isotropic surface energies is in the form of a doubly spherical lens. The ratio of the specific surface energies is

$$\sigma_{\alpha\alpha}/2\sigma_{\alpha\beta} = \cos\tfrac{1}{2}\theta.$$

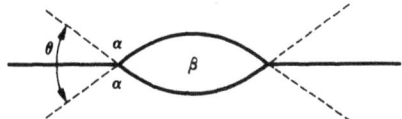

Fig. 1.7. Nucleation of β-phase in an isotropic $\alpha\alpha$ grain boundary.

The ratio of the critical energy of formation of a boundary nucleus ΔG_B^* to that for homogeneous nucleation of a spherical nucleus ΔG_H^* is

$$\Delta G_B^*/\Delta G_H^* = \tfrac{1}{2}(2 - 3\cos\theta + \cos^3\theta). \tag{1 8}$$

Because there are fewer potential sites available for grain-boundary nucleation than for homogeneous nucleation within the grains, the nucleation rate for a grain size d is changed from the homogeneous rate by a factor

$$\frac{\delta}{d}\exp\left(\frac{\Delta G_H^* - \Delta G_B^*}{kT}\right), \tag{1.9}$$

where δ is the grain-boundary thickness. The order of magnitude of the ratio δ/d is typically 10^{-6}, so by substituting this in (1.9), one would expect to observe grain-boundary nucleation being faster than homogeneous nucleation when $\Delta G_H^* - \Delta G_B^* \geqslant 14kT$.

Cahn (1956, 1957) has derived equations for the rates of heterogeneous nucleation at various singularities in a solid. On the basis of incoherent nuclei and isotropic surface tension, he has compared the barriers for nucleation on grain boundaries, corners and edges. Fig. 1.8 illustrates the conditions giving the highest volume nucleation rate for each kind of defect. In the perfect catalysis case (with $\sigma_{\alpha\alpha}/\sigma_{\alpha\beta} = 2$) nucleation can take place without the aid of the activation energy, and the rate of formation of the nucleus is then only determined by the activation energy of grain-boundary diffusion, U_b. From fig. 1.8 it is clear that for substrates of low $\sigma_{\alpha\alpha}/\sigma_{\alpha\beta}$ and high Δg_c, homogeneous nucleation dominates, whereas corners dominate only at the lowest Δg_c.

Observations by field-ion microscopy and by electron microscopy indicate that the structure of a curved boundary surface is not constant, but that imperfections such as ledges and grain-boundary dislocations are present in the interface. It is frequently observed that grain-boundary nucleation occurs only in certain regions of the boundary, e.g. in the case of the precipitation of $(Cr,Fe)_{23}C_6$ from stainless steel where the shape and density of grain-boundary particles vary with the orientation of the boundary. It appears probable therefore, that these heterogeneities of distribution of grain-boundary phases are associated with the presence of singularities in the grain-boundary structure itself.

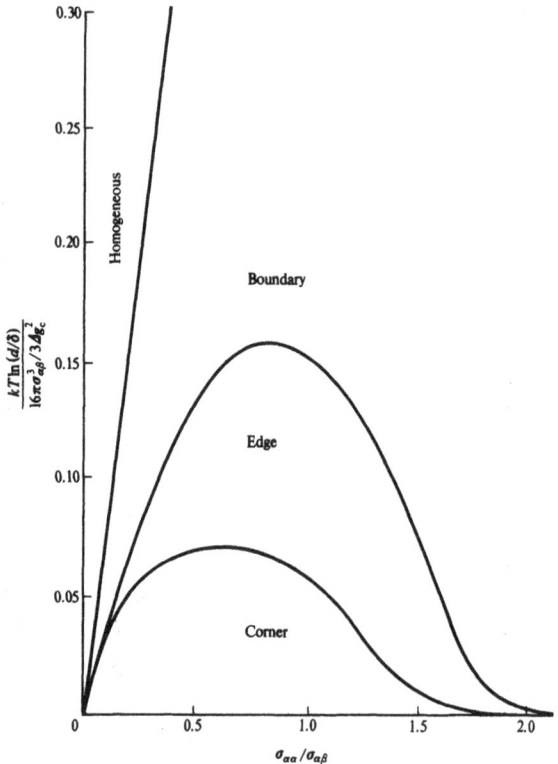

Fig. 1.8. Conditions giving the greatest volume nucleation rates for various kinds of defect. The ordinate represents the opposing factors $kT \ln (d/\delta)$ and the homogeneous nucleation energy, ΔG_H^*. The abscissa represents the catalytic effectiveness of the grain boundaries, with $\sigma_{\alpha\alpha}/\sigma_{\alpha\beta} = 0$ denoting zero catalytic effect (a contact angle of π) and $\sigma_{\alpha\alpha}/\sigma_{\alpha\beta} = 2$ giving perfect catalysis (After Cahn, 1956.)

Nucleation on other planar defects

Fig. 1.9 provides an example which compares the precipitation kinetics of a particular phase (the carbide $(Cr,Fe)_{23}C_6$) in association with various interfaces in an austenitic stainless steel. A family of 'C' curves is shown which represent the time of first appearance of the carbide at a particular site. In comparison with grain-boundary nucleation, it is seen that *twin boundaries* form nuclei at longer times, because of their lower interfacial energy (1.7). In accord with this principle, *coherent twin* interfaces are seen to be more difficult to nucleate upon than *incoherent twin* interfaces. On the other hand, it is seen that $\delta\gamma$-phase boundaries in the matrix are more effective than γ-grain boundaries in promoting carbide nucleation. This is because the energy of $\delta\gamma$-interfaces will be higher than

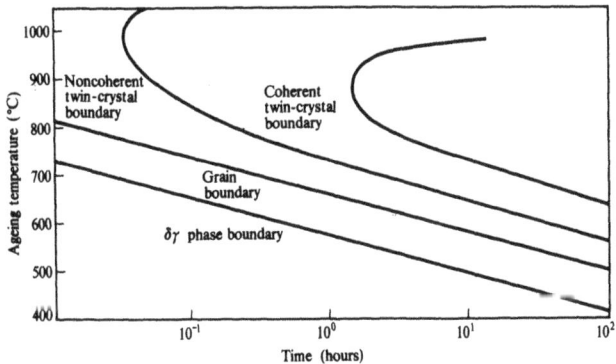

Fig. 1.9. Rate of nucleation of $(Cr, Fe)_{23}C_6$ at different types of interfaces in an austenitic stainless steel. (After Stickler & Vinckier, 1963.)

that of $\gamma\gamma$-boundaries due to a change in composition, as well as a change in structure across them.

Superlattice domain boundaries can act in a similar way in suitable systems: for example, in α-Fe-Al alloys there is evidence of domain boundaries catalysing nucleation of precipitates. There is also good evidence of precipitate nucleation at *stacking-faults*. The segregation of solute atoms to pre-existing stacking-faults, first proposed by Suzuki, may lower the stacking-fault energy and thus generate zones within the crystal which differ in structure and composition from the matrix. If a phase can be formed with the same stacking sequence as the fault, then its nucleation will be promoted by the contribution to surface-energy or strain-energy terms in (1.7). The Al-Ag system illustrates this effect well, in that a hexagonal γ'-phase forms on the $\{111\}$ planes of the fcc matrix.

1.2.3   Strain-energy effects

Lattice misfit between a precipitate nucleus and the matrix can arise from the difference in lattice parameters or the difference in lattice symmetry of the two phases. As illustrated in figs. 1.4a and b, the misfit can be accommodated by elastic distortions or by defects lying in the interface.

If the unit cell sides of the precipitate and matrix are a_β and a_α respectively, then the misfit between the lattices (δ) is defined as

$$\delta = \frac{2(a_\beta - a_\alpha)}{a_\beta + a_\alpha} \approx \frac{a_\beta - a_\alpha}{a_\beta}. \tag{1.10}$$

In order to calculate g_e, Nabarro (1940) assumed that the nucleus *and* matrix are both strained isotropically, and applied a continuum mechanics

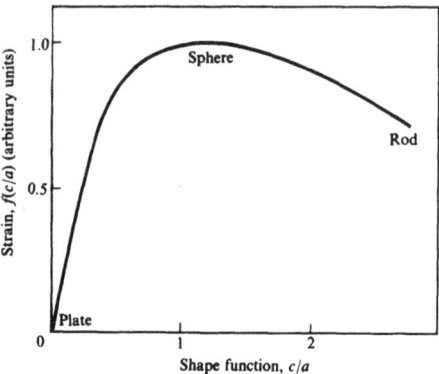

Fig. 1.10. The strain, $f(c/a)$, of an incoherent spheroidal nucleus as a function of its shape: a is the radius and $2c$ the thickness of the spheroid. (After Nabarro, 1940.)

approach to obtain the value of g_e, which was *independent* of the shape of the particle

$$g_e = \frac{6G_\beta \delta^2}{1 + 4G_\beta/3K_\alpha},$$ (1.11)

where G_β is the shear modulus of the precipitate phase and K_α is the bulk modulus of the matrix. If the value of Poisson's ratio is taken as 0.3 then $K = 2G$, so if the moduli of the matrix and precipitate are similar, then

$$g_e \approx 4G\delta^2.$$

For hard particles in a softer matrix, it is more realistic to assume that the nuclei are unstrained, and that only the matrix suffers hydrostatic strain. In this case Nabarro found that g_e becomes a function of the shape of the particle, in that the specific strain-energy decreases in the direction sphere → rod → plate (fig. 1.10).

Plate-shaped nuclei are frequently observed in systems of practical importance. There are factors additional to that of maintaining g_e which favour such a particle shape, namely: (i) the anisotropy of the *surface energy*, so that the habit plane of the plate is that of best crystallographic fit; (ii) The anisotropy of the *elastic constants* of the crystal lattice, so that the habit plane of the plate is perpendicular to the direction of minimum Young's modulus; and (iii) the planes in which dislocations dissociate, or upon which vacancy clusters condense may lead, as already discussed, to the nucleation of plate-like precipitate particles.

In describing the structural relationship between the matrix and a precipitate there are two independent factors which must be considered. Firstly, the *orientation relationship* between the crystal lattices of matrix and precipitate must be described. This is normally defined by planes and/or directions in each phase which are parallel, e.g.

$$(h_1 k_1 l_1)_\alpha \parallel (h_2 k_2 l_2)_\beta, \quad [U_1 V_1 W_1]_\alpha \parallel [U_2 V_2 W_2]_\beta.$$

Clearly, two sets of parallel planes or two sets of parallel directions in each phase would be equally suitable.

Secondly, in the case of plate-like precipitates, the *habit plane* must be described. This is the plane of the matrix which is parallel to the plane of the precipitate plate. Obviously this parameter does not arise in the case of spherical particles, and in the case of *rod-shaped* precipitates, only the matrix direction parallel to the rod axis need be defined.

When a *semicoherent* nucleus is considered, some of the elastic strain-energy associated with the nucleus is replaced by interfacial dislocations (fig. 1.4b) which contribute to the interfacial energy, σ. If a nucleus of n unit cells of each lattice is now considered, the misfit m is given by

$$m = n(a_\beta - a_\alpha).$$

By introducing an interfacial dislocation with a component of Burgers vector b_m in the direction of misfit defined above, the effective misfit, m_{eff}, is reduced to

$$m_{\text{eff}} = n(a_\beta - a_\alpha) - b_m$$

and so

$$\delta = \frac{n(a_\beta - a_\alpha) - b_m}{n\, a_\beta}. \tag{1.12}$$

Thus $\delta = 0$ when $n(a_\beta - a_\alpha) = b_m$, and only the interfacial energy term need be considered in (1.7).

1.2.4   The formation of transition phases

When supersaturated solid solutions decompose, one or more metastable *transition phases* may appear prior to, or in addition to the equilibrium precipitate. By definition, metastable phases possess a smaller negative Δg_c at a given temperature than the more stable phase, so that if common tangents to free-energy/composition curves are envisaged it will be obvious that metastable phases are in metastable equilibrium with a matrix of different (higher) solute concentration than the stable phase. Consequently, there will be as many metastable phase diagrams as there are metastable phases.

This situation is shown for the three phases θ'', θ' and θ which form in Al–Cu solid solutions, in fig. 11a, and the crystallographic features of these phases are indicated in fig. 1.11b. It should perhaps be emphasized that the *prediction* of the number of transition phases, their crystal structure and composition has not yet been achieved for any alloy system, even on an empirical basis.

In considering the *nucleation sequence* of transition phases, we will discuss the particular case of the Al–Cu system, although the conclusions are of quite general applicability. In that system, if the supersaturated solid solution is aged at temperatures below all the solvi of fig. 1.11a, the experimentally observed ageing sequence is

Supersaturated α solid solution $\rightarrow$ GP zones $\rightarrow \theta'' \rightarrow \theta' \rightarrow \theta$.

Following Russell & Aaronson (1975), we will first consider this sequence in terms of the Δg_c term in (1.7). Assuming that the supersaturated solid solution is sufficiently dilute to allow replacement of activities by compositions and that the precipitate phase (β) is markedly solute-rich,

$$\Delta g_c = \frac{RT}{V_\beta} \ln \frac{x\alpha^{\alpha\beta}}{x\alpha} \, , \tag{1.13}$$

where R is the gas constant, V_β the molar volume of the β-(precipitate)

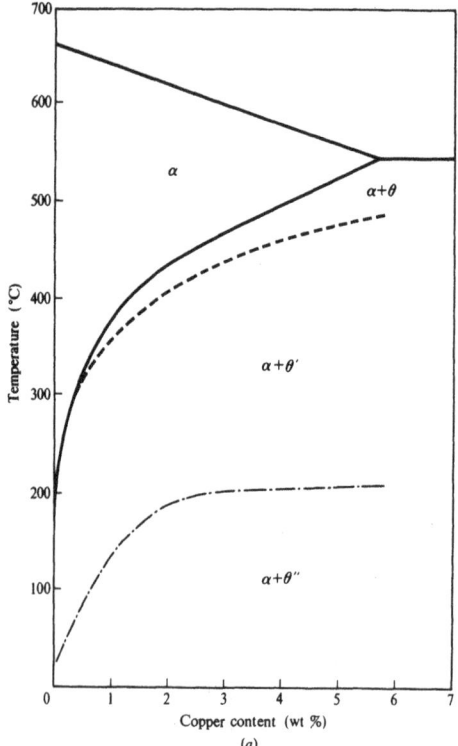

(a)

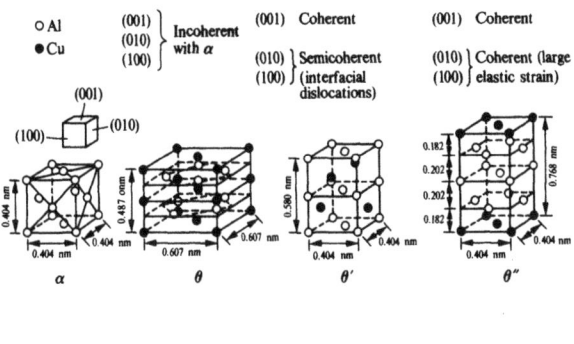

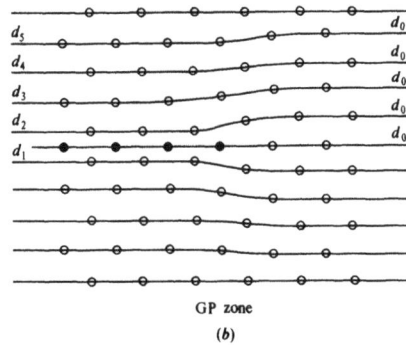

Fig. 1.11. (*a*) Metastable equilibrium solvus curves for Al-rich Al-Cu alloys. (*b*) Equilibrium-phase (θ) and transition-phase (θ', θ'', GP zone) crystal structures in Al rich Al-Cu alloys. (After Hornbogen, 1967.)

phase, x_α the mole fraction of solute in the matrix prior to ageing, and $x_\alpha^{\alpha\beta}$ the mole fraction of solute in the α-phase at the $\alpha/(\alpha + \beta)$ solvus. It follows from fig. 1.11*a* that the nucleation driving force, Δg_c, is successively more negative (at the same temperature and x_α) for GP zones θ'', θ' and θ (assuming that all the phases have substantially the same composition). If Δg_c were the primary factor controlling the kinetics of nucleation at a given temperature, the θ-phase would then be the first to nucleate, rather than the last. Another factor, that of interfacial energy, must therefore be governing the sequence of precipitation.

The interfacial energy barrier decreases with progression through the series θ, θ', θ'' and GP zones. Therefore, once Δg_c becomes sufficiently negative for all possible precipitates, the interfacial energy factor is dominant, since ΔG^* in (1.5) is proportional to σ^3, but only to $(\Delta g_c + g_e)^2$, and this can account for the observed sequence of precipitates.

A quantitative treatment of this problem requires determination of the equilibrium shape for each precipitate involved, at the type of site at which it nucleates. The energy of each bounding face would then have to

Table 1.1 *Combination of crystallographic conditions and defects.* ΔG^* *increases in the direction of the arrow (for* Δg_c *nv constant).* (After Hornbogen, 1969.)

Crystallographic relationship of α and β	Grain boundary (B)	Dislocation (D)	Vacancy (V)	Homogeneous nucleation in perfect lattice (H)
Coherent (c)	Bc $\rightarrow$	Dc $\rightarrow$	Vc $\approx$	Hc
	$\downarrow$	$\downarrow$	$\uparrow$	$\uparrow$
Semicoherent (s)	Bs $\rightarrow$	Ds $\leftarrow$	Vs $\leftarrow$	Hs
	$\downarrow$	$\uparrow$	$\uparrow$	$\uparrow$
Noncoherent (n)	Bn $\leftarrow$	Dn $\leftarrow$	Vn $\leftarrow$	Hn

be calculated from the sum of the structural component of energy (associated with interfacial misfit dislocations) and the chemical component (resulting from a different proportion of atoms of each species surrounding atoms in the boundary, compared with atoms within each crystal).

In table 1.1 a scheme is given for the *combination* of certain defects in the matrix (grain boundaries, dislocations and vacancies) with the three crystallographic conditions of precipitation, namely coherent, semi-coherent and noncoherent. A fourth column in the table considers homogeneous nucleation of these precipitates in a perfect lattice. If it is assumed that Δg_c *nv* has the same value for all combinations (1.1), then ΔG^* (1.3) can be expected to decrease in the direction of the arrows, due to the interaction of the lattice defect and the nucleus. From this approach it is evident that four types of nucleation are associated with a minimum value of ΔG^*.

(i) *Noncoherent precipitates upon grain boundaries.* In this case the grain-boundary energy is effective in decreasing the surface-energy term, σ, of the activation energy for nucleation (1.5). The formation of a coherent phase at a grain boundary is less probable, since the nucleus fits better into a perfect lattice, and the grain-boundary energy does not lower the surface-energy term.

(ii) *Semicoherent precipitates upon dislocations.* Since such particles contain interfacial dislocations for lattice matching, a pre-existing dislocation or a dislocation that can be formed easily by a dislocation reaction will again lower the activation barrier.

(iii) *Coherent precipitates upon vacancies.* This can arise if, for example, the specific volume of the nucleus is larger than that of the matrix.

(iv) *Coherent phases homogeneously nucleated.* This is most likely if the nucleus fits without elastic distortion, so that process (iii) above is not favoured.

1.2.5   Nucleation of one phase at different types of defect

Experimental observations suggest that the sequence of nucleation at grain boundaries, dislocations and in the matrix can be quite different in different alloys.

For noncoherent nuclei, the most frequently observed sequence is that corresponding to table 1.1, namely the relative rates of nucleation are

$$\text{grain boundaries} > \text{dislocations} > \text{matrix.}$$

For semicoherent nuclei the rates

$$\text{dislocations} > \text{matrix} > \text{grain boundaries,}$$

or

$$\text{dislocations} > \text{grain boundaries} > \text{matrix,}$$

are expected.

For coherent particles with a stress field, the most frequently observed sequence is

$$\text{dislocations} > \text{matrix} > \text{grain boundaries,}$$

providing that the misfit is so large that grain boundaries become the most preferred sites. Coherent nuclei with zero misfit show the sequence

$$\text{matrix} > \text{dislocations} > \text{grain boundaries.}$$

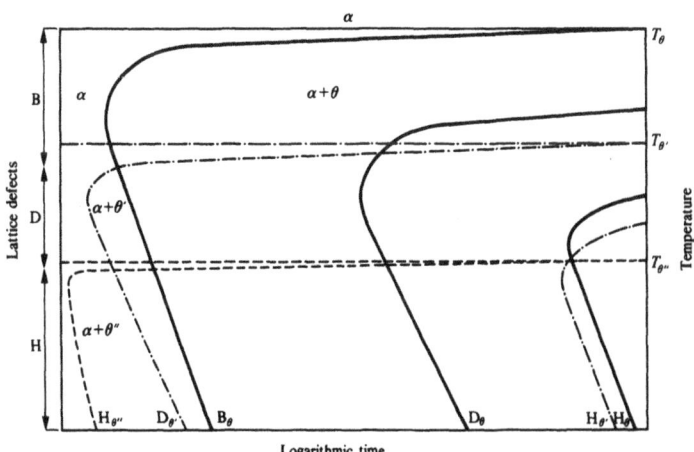

Fig. 1.12. Schematic nucleation diagram showing a primary phase (α) plus three precipitation phases, θ, θ' and θ'', and several defects as parameters. B refers to nucleation on grain boundaries, D to nucleation on dislocations and H to nucleation in a perfect lattice or at vacancy clusters. (After Hornbogen, 1969.)

1.2.6   Nucleation of several phases at different types of defect

The Al–Cu system will be used once more to illustrate this more complex situation. We will confine our attention to three precipitate phases, the noncoherent stable phase θ, the semicoherent θ' and the coherent θ''. We will not consider the coherent GP1 zones separately, since they transform *in situ* to θ'' as ageing progresses.

The situation can be summarized on a schematic temperature–time constant-composition section of a nucleation diagram, fig. 1.12. The differing nucleation rates given by (1.7) for the various phases indicated in the phase diagram of fig. 1.11a give rise to the '*C*' curves shown. U_D can be regarded as temperature-independent only to a first approximation, since there is an anomalous increase in diffusivity at low temperatures due to the increased supersaturation of vacancies.

In the lowest temperature range (fig. 1.12) the first phase to form is θ'' in the matrix. This is followed by θ'-particles at dislocations and θ-particles in the grain boundaries. As the particles continue to grow, their elastic coherency stress increases, and when this stress reaches the shear strength of the matrix the coherency will break down and a dislocation ring will form (fig 1.6). At this dislocation, a θ'-particle can nucleate. Being more stable, such θ'-particles will grow at the expense of the θ''-phase.

As the θ'-particles grow, an increasing number of interfacial dislocations form that finally reach such a density that portions of the interfaces of the θ'-particles become noncoherent, so a θ-nucleus begins to form. Again, the stable phase will grow at the expense of the metastable phases present, so that after prolonged ageing only the θ-phase will be present, as predicted by the equilibrium diagram (as opposed to the metastable *phase* diagram of fig. 1.11a).

Russell & Aaronson (1975) have pointed out that when the first transition phase to form makes its appearance, at least two depressant effects are exerted upon the nucleation probability of any succeeding phases. First of all, Δg_c for nucleation of the other phases becomes less negative at all nucleation sites lying within the (time-dependent) diffusion fields of the individual precipitates of the first phase. Further, nucleation of the first phase consumes some porportion of the available nucleation sites. This may be particularly important if the initial phase forms on some easily saturable sites, such as dislocations.

Given these handicaps, which the nucleation and growth of the first transition precipitate impose upon the nucleation of subsequent precipitates, Russell & Aaronson suggest that the nucleation of the next precipitate should occur preferentially at the interphase boundaries of the first one, and that succeeding precipitates ought to follow the same pattern. This will occur because of the greater importance of minimizing

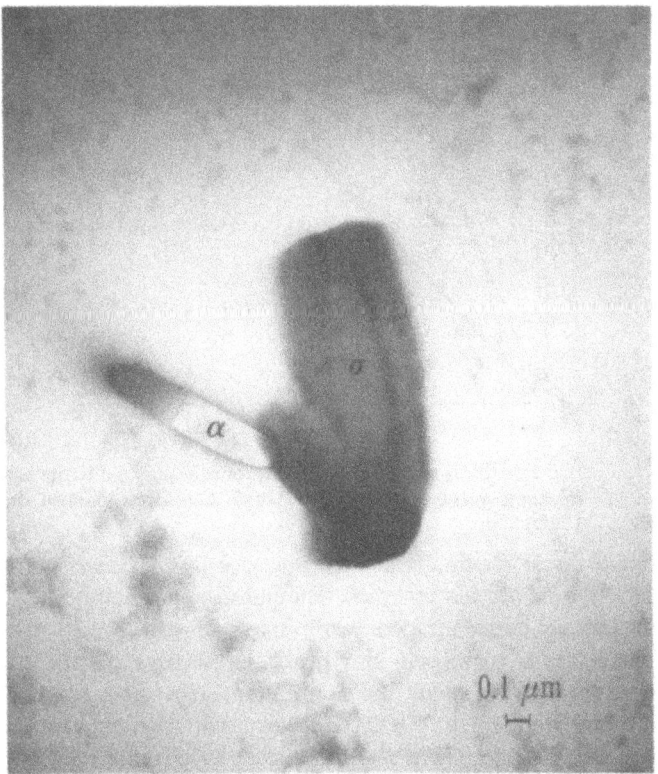

Fig. 1.13. Electron micrograph showing the nucleation and growth of
σ-phase upon ferrite (α-phase) in an austenitic stainless steel aged 1500
hours at 750 °C. (Courtesy of L. K. Singhal.)

the interfacial free energy than of Δg_c, and also because the interphase
boundaries, being two-dimensional features, can usually contribute much
more energy to the nucleation process than dislocations.

Transmission electron microscopy has in fact demonstrated that in
numerous systems of practical interest one precipitate phase has nucleated
at the interphase boundaries of another. For example, in the Al–Cu system
there is evidence for the nucleation of the θ-phase at θ', and of θ' at θ''.
Fig. 1.13 illustrates the effect in an austenitic stainless steel, where the
σ-phase (which is an embrittling phase, often sluggish to nucleate) is seen
to have nucleated upon a pre-existing body-centred cubic (bcc) ferrite
particle.

1.2.7.   Precipitate growth from supersaturated solid solutions

In the analysis of precipitate growth from supersaturated solid solutions,
the possible limiting factors considered are the rate at which atoms are

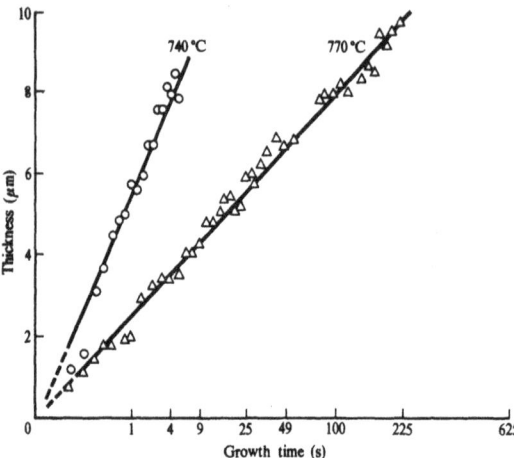

Fig. 1.14. Plots at 740 °C and 770 °C of pre-eutectoid ferrite allotriomorph thickness versus time in Fe–0.11wt% C. (After Kinsman & Aaronson, 1967.)

brought to or removed from the interface by *diffusion*, and the rate at which they *cross the interface*. The interface reaction is likely to be the rate-limiting step during the early stages of growth since the diffusion distance tends to zero in this situation. At large particle sizes, lattice diffusion is likely to be the slower step, since the continuous removal of solute from solution reduces the concentration gradient (the driving force for diffusion). We shall only refer to diffusion-controlled growth here, since this is the process which has been analysed in greater detail.

The subject has been recently reviewed in detail by Martin & Doherty (1976), and we will here simply refer to the calculations of Aaron, Fainstein & Kotler (1970). These workers predicted that the growth of spherical precipitates should obey an equation of the form

$$r = \alpha \, (Dt)^{\frac{1}{2}}, \tag{1.14}$$

where r is the particle radius after time t, D (assumed independent of composition) is the volume diffusion coefficient and α is a function of the supersaturation. A similar calculation made for the growth of planar precipitates again predicted a parabolic relationship between thickness and time.

Most of the data available to test the above theories relate to the formation of crystals of pre-eutectoid ferrite forming in austenite grain boundaries in steel (grain-boundary allotriomorphs). Typical data are shown in fig. 1.14 where it is seen that thickening of the particles obeys a parabolic law in agreement with (1.14).

Rather less good agreement is obtained for the growth of Widmannstätten ferrite plates growing wholly within the interior of a parent grain, where the observed rates of thickening are less than the theoretically predicted values. This may be due to the good crystallographic fit between particle and matrix in such structures leading to low interface mobility.

Precipitate growth in saturated solid solutions

When precipitation from supersaturated solid solution is complete, further annealing leads to precipitate coarsening driven by the interfacial free energy between the precipitate and the matrix - the process known as *Ostwald ripening*. The physical process by which the microstructure coarsens and releases its excess surface energy is due to the higher solubility of small particles, since these have a larger ratio of surface area to volume. The larger particles thus grow at the expense of the smaller ones, and the growth kinetics depend upon the rate-controlling step in the process.

If *lattice diffusion* is the rate-controlling step, then the analyses of Wagner (1961) and Lifshitz & Slyozov (1961) lead to a coarsening rate given by the expression

$$\bar{r}_t^3 - \bar{r}_0^2 = 8D\sigma V_m C_\alpha(\infty)t/9RT, \tag{1.15}$$

where

$\bar{r}_t$	is the average particle radius at time t,
$\bar{r}_0$	is the average particle radius at time $t = 0$,
D	is the volume diffusion coefficient of the diffusing species,
σ	is the specific surface free energy of the particle/matrix interface,
V_m	is the molar volume of the precipitate,
$C_\alpha (\infty)$	is the equilibrium solute content in the matrix,
R	is the gas constant,
T	is the temperature in K.

When the rate of coarsening is controlled by the rate of atom transfer across the *interface*, Wagner predicts:

$$\bar{r}_t^2 - \bar{r}_0^2 = 64KV_m C_\alpha (\infty)t/81RT, \tag{1.16}$$

where K is a proportionality constant that includes the interface mobility.

Finally, if the coarsening precipitates are situated on grain boundaries and the dominant diffusion path is that down the grain boundaries, Kirchner (1971) obtains

$$\bar{r}_t^4 - \bar{r}_0^4 = 9\delta D_B \sigma C_\alpha (\infty)_B V_m t/32ABRT, \tag{1.17}$$

where

δ	is the grain-boundary thickness,

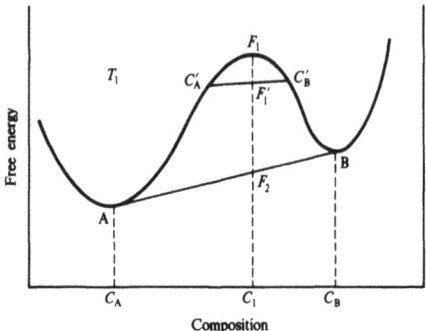

Fig. 1.15. Free-energy/composition relationship corresponding to spinodal decomposition.

D_B is the grain-boundary diffusion coefficient,

$C_\alpha (\infty)_B$ is the solute content at a grain boundary in equilibrium with an infinitely large precipitate,

A is the parameter defined by

$$A = \tfrac{2}{3} - (\sigma_B/2\sigma) + \tfrac{1}{3}(\sigma_B/2\sigma),$$

B is the parameter defined by

$$B = \tfrac{1}{2} \ln (1/f),$$

σ_B is the grain-boundary energy,

f is the fraction of the grain boundary covered by precipitates.

Experimentally it is found that in the vast majority of cases the kinetics are diffusion-controlled, that is, the mobility of the interface does not limit the reaction.

1.3   Precipitate distribution

1.3.1   Homogeneous distribution

Precipitation by *spinodal decomposition* is a situation which leads to a very uniform, homogeneous distribution of coherent second-phase particles. In this type of system, no structural change is involved in the transformation, but compositional changes do occur and fig. 1.15 shows a free-energy/composition curve for a homogeneous phase which is meta-stable with respect to phases A and B between compositions C_A and C_B.

An alloy of initial composition C_1 and free energy F_1 decomposes into a mixture of two phases of composition C_A and C_B and average free energy F_2. At an early stage in the transformation the compositions of the two separating phases may be C'_A and C'_B with an average free energy F'_1

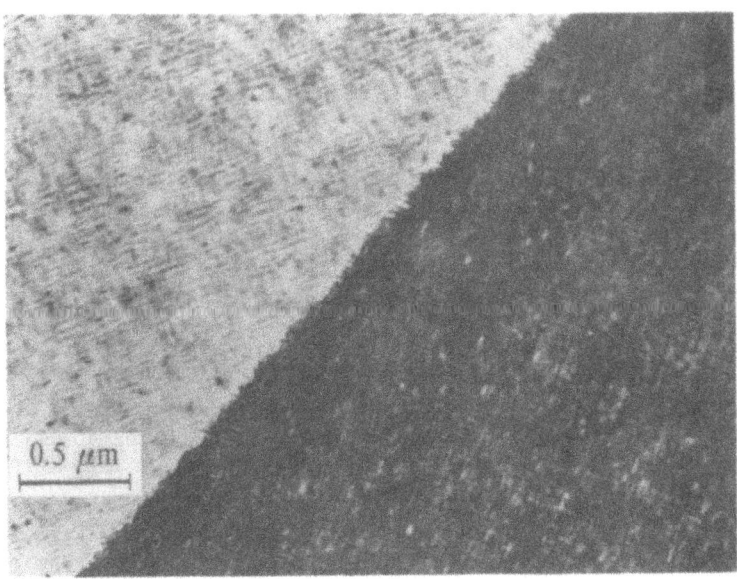

Fig. 1.16. Electron micrograph showing the uniformity of spinodal decomposition in a Cu–Ni–Fe alloy.

which is always lower than F_1, so the composition proceeds with a *continual* decrease in free energy and there is no thermodynamic barrier to decomposition of the solid solution. Spinodal decomposition thus occurs simultaneously throughout the matrix, although there is no precise stage at which the new phases appear.

The Cu–Ni–Fe system exhibits spinodal decomposition, and the transformed structure consists of regularly spaced quasi-spherical particles at small volume fractions and a lattice of interconnected rods lying along ⟨100⟩ directions at large volume fractions (fig. 1.16). It is striking that there is no preferential precipitation on grain boundaries or other lattice defects. Since such nucleation arises (in the absence of spinodal decomposition) because of the reduced activation energy for nucleation at such defects, such an effect would not be expected in spinodal alloys where no nucleation barrier to precipitation exists.

Theories of the transformation take into account the positive free-energy contribution arising from the bond energies and elastic strains across the diffuse interface between phases. In the treatment due to Cahn & Hilliard (1958, 1959) the free energy of the material is given by the expression

$$F = \int_V [F'(C) + K(\nabla C)^2 + \eta^2 E(C - C_0)^2/(1 - \nu)]\ dV, \quad (1.18)$$

where

$F'(C)$	is the Helmholtz free energy per unit volume of homogeneous material of composition C,
$K(\nabla C)^2$	is a measure of the increase in energy due to the non-uniform environment of atoms in a concentration gradient,
$\eta^2 E(C - C_0)^2/(1 - \nu)$	is a measure of the strain-energy associated with the interface,
E	is Young's modulus,
ν	is Poisson's ratio,
C_0	is the average composition of the solid solution,
η	is the linear expansion of the lattice per unit composition change.

In the spinodally decomposed microstructure, assuming that C varies sinusoidally with distance x, so that

$$C - C_0 = B \cos \beta x,$$

the wavelength of sinusoidal fluctuations will be given by $2\pi/\beta$. It is found that the alloy is unstable to fluctuations above a critical wavelength, which depends on the supersaturation, and is given by

$$\frac{2\pi}{\beta_c} = \left[\frac{-8\,\pi^2 K}{\mathrm{d}^2 F'/\mathrm{d}C^2 + 2\eta^2 E/(1 - \nu)}\right]^{\frac{1}{2}} = \lambda_c.$$

Provided, therefore, that $\mathrm{d}^2 F'/\mathrm{d}C^2$ is sufficiently negative (in the curve shown in fig. 1.15), the solution is unstable with respect to all (very small) composition fluctuations of a wavelength greater than λ_c. One might therefore expect that rather a complex microstructure might result, since this spectrum of wavelengths should be present in all directions in the material. The kinetics of the decomposition simplify the situation, however. Although the driving force is greatest for the largest wavelength, the diffusion distance is least for the smallest stable wavelength. These two factors lead to the conclusion that the fastest growing wavelength has a value $\lambda_c \sqrt{2}$, and this is the one observed in metallographic study (although the progress of the full spectrum can be followed by X-ray diffraction techniques). Elastic anisotropy means that the alloy is first unstable to fluctuations in a particular direction where E is a minimum: this direction is usually $\langle 100 \rangle$ in cubic materials (see fig. 1.16).

Unfortunately, the stronger spinodal structures are very brittle, failing in an intergranular manner. The real potential of this type of structure remains to be exploited, and most precipitated microstructures of practical importance are developed by nucleation-and-growth processes and hence

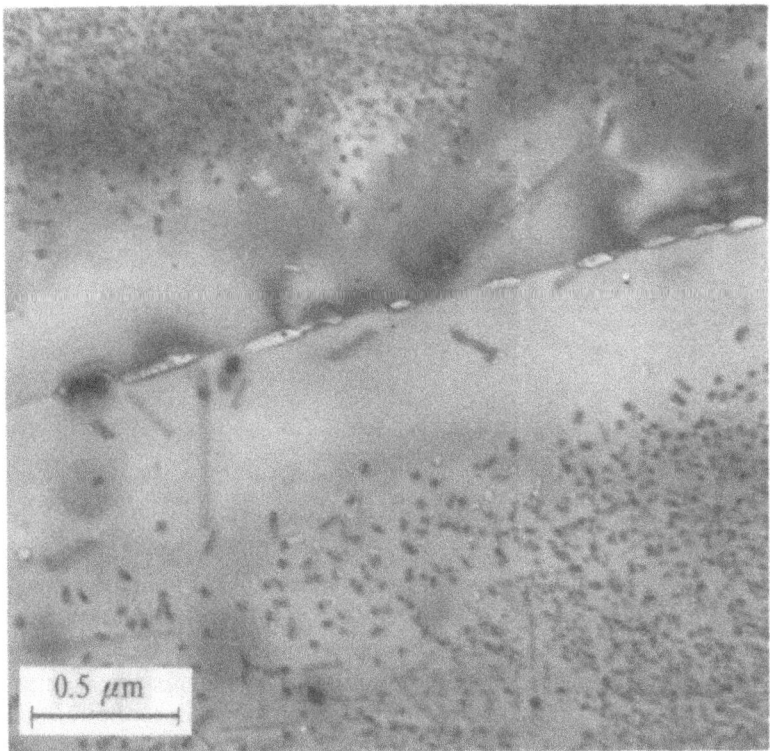

Fig. 1.17. Electron micrograph showing the precipitate-free zone adjacent to a grain boundary in Al-4wt%Zn-3wt%Mg aged for 1 day at 150 °C. (Courtesy of I. J. Polmear.)

exhibit a heterogeneity of particle distribution due to the influence of the crystal defects we have discussed earlier.

1.3.2   Heterogeneous distribution

We will consider in turn the influence of three types of defect found in metal matrices upon the distribution of precipitates in quenched and aged alloys.

Grain boundaries

Aged alloys usually exhibit zones denuded of precipitate adjacent to the grain boundaries (fig. 1.17). There are two ways in which such precipitate-free zones (PFZ) can arise: either from a local depletion of solute atoms there; or from a depletion of lattice vacancies which, when the alloy is quenched, enter the 'sink' provided by the grain boundary. We will consider initially the latter factor, taking as an example the formation of

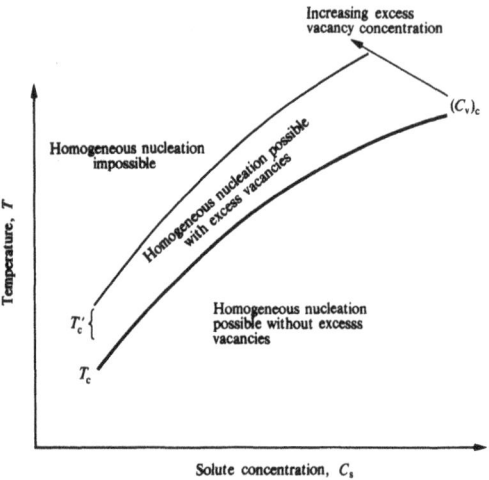

Fig. 1.18. Schematic diagram showing the variation of T and T'_c with solute concentration. (After Jacobs & Pashley, 1969.)

PFZs in aluminium alloys aged to precipitate GP zones within the grains.

Jacobs & Pashley (1969) define (see fig. 1.18) a critical temperature at which homogeneous nucleation can occur when no excess vacancies are present at the ageing temperature. Homogeneous nucleation does not occur above this temperature because of inadequatè solute supersaturation. This temperature, T_c, decreases as the solute concentration decreases (fig. 1.18), in a similar manner to the GP zone solvus line (fig. 1.11a). Jacobs & Pashley further postulate that nucleation is possible with a given critical excess-vacancy concentration $(C_v)_c$ up to a maximum temperature T'_c (fig. 1.18). An excess of vacancies will arise when the specimen is quenched from the temperature of solution heat-treatment. The higher number of vacancies in equilibrium with the lattice at the upper temperature will tend to be retained in supersaturation at the lower temperature. There are two ways in which these excess vacancies may influence homogeneous precipitation above temperature T_c; either by their contribution to the diffusivity of solute which accelerates the formation of clusters during the quench, and which then give rise to the intermediate precipitate; or by providing vacancy aggregates which nucleate a new phase.

During, and for a short period after a quench from the solution-treating temperature to the ageing temperature (T_A), an excess-vacancy profile will be established in the vicinity of each grain boundary (fig. 1.19). This is because grain boundaries are sinks for vacancies, and the shape of the excess-vacancy profile will depend on the solution-treatment temperature, since the equilibrium concentration of vacancies increases with temperature

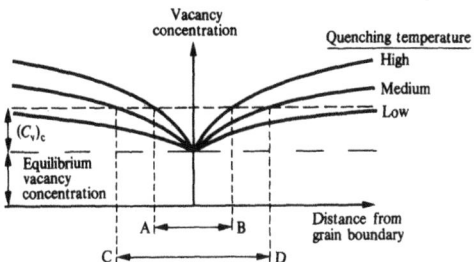

Fig. 1.19. Excess vacancy profiles near to a grain boundary. (After Jacobs & Pashley, 1969.)

and also with the quenching rate. Slower rates will allow diffusion over larger distances and hence result in a decreased slope in the profile. Fig. 1.19 shows the profiles for a specimen quenched from a high, a medium and a low solution-treatment temperature. If the ageing temperature, T_A, is above T_c, nucleation of solute clusters can occur in regions of the specimen where the excess-vacancy concentration is greater than the critical value, $(C_v)_c$. In fig. 1.19, AB thus defines the vacancy-depleted PFZ in the sample quenched from the highest solution-treatment temperature. Since, in this example, the vacancy profile established on quenching from the lowest solution-treatment temperature is shown as lying below $(C_v)_c$, homogeneous nucleation is not possible in this case, and only heterogeneously nucleated precipitates can occur.

The width of the PFZ also decreases as the ageing temperature is decreased. Under these circumstances the greater solute supersaturation allows a smaller critical nucleus size, thereby reducing the critical vacancy concentration required for nucleation.

Solute-depleted PFZs will arise when solute is lost from the vicinity of the boundary during the quench, either because precipitates are nucleated heterogeneously on the boundary during cooling to T_A, or because preferential segregation of solute occurs at the boundary during cooling to the ageing temperature. In either case a solute-concentration profile will be set up as indicated in fig. 1.20, whose form will depend upon the structure and misorientation of the boundary itself, owing to the influence of these parameters on the nucleation of grain-boundary precipitates. Reference to fig. 1.20 indicates that with a certain excess-vacancy concentration at the ageing temperature there is a critical solute concentration C_s required for nucleation to occur. Fig. 1.20 illustrates the values of C_s for various temperatures as horizontal lines, and the intersection of these lines with the solute-concentration profile gives the width of the solute-depleted PFZ, which is seen to decrease from YY to XX as the ageing temperature is decreased from 150 °C to 140 °C. Only solute-depleted

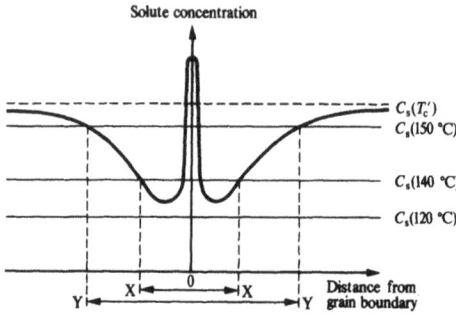

Fig. 1.20. Schematic diagram of the solute-concentration profile after quenching. (After Jacobs & Pashley, 1969.)

PFZs are possible when ageing below T_c, and above this temperature a vacancy-concentration profile must be superimposed on the solute-concentration profile. The PFZs which form above T_c may have widths of up to 1 μm or more, whereas solute-depleted PFZs formed below this temperature are usually narrower than 0.2 μm.

Two-step ageing treatments and PFZs

If an age-hardenable alloy is rapidly quenched to a temperature above T_c, and aged at this temperature, the microstructure will be similar to that illustrated in fig. 1.17. If this alloy is then held *below* T_c for a period of time and then aged again above this temperature, a finer precipitate is formed within the original vacancy-depleted PFZ. Jacobs & Pashley have also shown that quenching directly to an ageing temperature *below* T_c and holding for various times has a marked effect on the subsequent PFZ formed above T_c (fig. 1.21).

These phenomena can be explained by assuming that clusters are formed below t_c, and if the sample is subsequently aged at a higher temperature, these clusters develop into the intermediate precipitate, or dissolve, depending on whether they are larger or smaller than r^*, the critical size at that temperature. Since excess vacancies are not required for nucleation *below* T_c, the clusters form right up to the grain boundary if there is sufficient solute. Since there is an excess-vacancy profile near the boundary, and since the solute diffusion rate depends on the vacancy concentration, clusters nucleated near the grain boundary will grow less rapidly than those nucleated further away. The width of the PFZ present after the sample is aged the second time above T_c depends on the holding time below T_c, since the probability of reaching the critical size for nucleation increases with time.

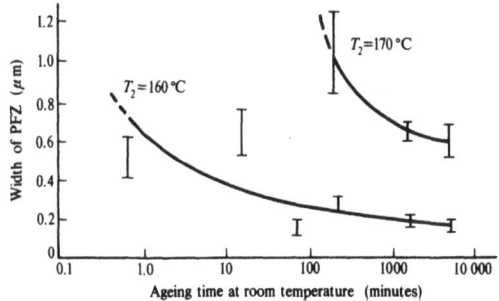

Fig. 1.21. The effect of pre-ageing time at room temperature on the width of the PFZ established on subsequent ageing at either 160 °C for 1 hour or 120 °C for 20 minutes. (After Jacobs & Pashley, 1969.)

Discontinuous precipitation

A fairly common way for precipitation to occur in a terminal solid solution which is undercooled several hundred degrees below the temperature of maximum solubility is *discontinuous precipitation*. In this case, nodules of a lamellar structure form at the grain boundaries and grow into the grains. The lamellae consist of plates or laths of precipitate embedded in the matrix, and the structure can form when the temperature is so low that nucleation and growth of individual precipitate particles within the matrix occurs at a negligible rate. The precipitate only forms at the advancing interface, so that, in a given region, precipitation occurs discontinuously with time instead of by the nucleation and continuous growth of particles.

The barrier to nucleation of the new phase is reduced because of its formation at a grain boundary. The required diffusion of solute takes place along the grain boundary and thus, since it involves a lower activation energy, will tend to be the dominant diffusion mechanism at relatively low temperatures. The energy barrier to nucleation will be further reduced if the new phase has a special orientation so that the energy of the $\alpha\beta$ interface is low between *one* of the grains forming the grain boundary and the precipitate. Fig. 1.22 illustrates successive stages in the start of discontinuous precipitation, which illustrates how the grain boundary moves as the nodule, or 'cell' of precipitate grows.

Because of the low nucleation rate, this process is unlikely to lead to finely-dispersed precipitate particles. Somewhat coarse, heterogeneous particle distributions are normally observed, therefore this method of decomposition is not one which is to be desired if strong microstructures are required. Meyrick (1976) has recently advanced an explanation for the observation of discontinuous precipitation in some alloy systems, but not in others. He suggests that it will only be observed in those alloys in which the solute is strongly absorbed at the grain boundaries. Meyrick points out

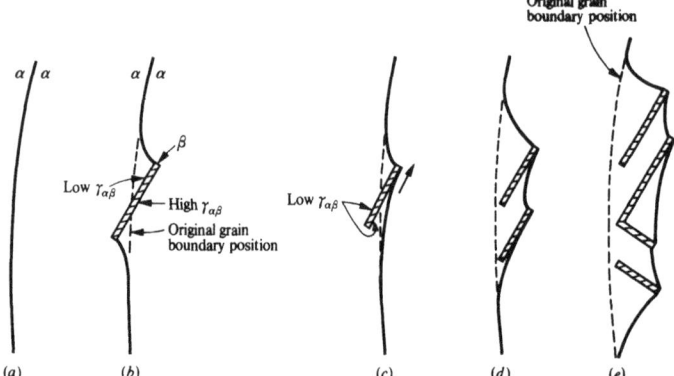

Fig. 1.22. Successive stages in the start of precipitation of β and α by discontinuous precipitation. (After Tu & Turnbull, 1967.)

that the total free-energy change associated with the onset of discontinuous precipitation will include a positive term corresponding to the increase in the effective grain-boundary energy due to the depletion of the solute population. The boundary can replenish its solute concentration by bowing into the adjacent matrix and thus reduce its energy. Provided that the total reduction exceeds the increase in energy caused by the increased grain-boundary area (as in fig. 1.22), bowing is energetically favoured.

Dislocations

Transgranular soft zones, through precipitate denudation, are commonly observed in aged alloys. The effect arises from the presence of matrix dislocations. For example, Al–Cu alloys may, on ageing, develop arrays of θ'-precipitates in association with the matrix dislocations, whereas the bulk of the matrix contains a fine dispersion of the coherent θ''-phase. The larger plates of the θ'-phase give rise to a local depletion of the θ''-precipitate density.

The dislocations giving rise to heterogeneous precipitate distribution can be formed by quenching stresses during quenching from the solution heat-treatment temperature. Such stresses can arise from differential thermal contraction in the alloy when specimens of appreciable cross-sectional area are quenched, such that steep temperature gradients are formed in the piece. The resulting dislocation arrays are known as *quench bands*.

Inclusions

Second-phase particles present in a supersaturated solid solution can act in two ways to perturb the uniform precipitation of an ageing precipitate.

Firstly, they may generate matrix dislocations because of quenching stresses arising from differential thermal contraction effects. These dislocations will in turn act as heterogeneous nucleation sites for the ageing precipitate. Secondly, however, the interface between the inclusion and the matrix may itself act as a preferential precipitation site during ageing.

The presence of heterogeneities forming substrates for phase nucleation can lead to the nucleation of undesired phases during the actual quenching process itself. This obviously reduces the supersaturation of the matrix to a certain extent, hence reducing the subsequent ageing response of the alloy. The tendency for an age-hardening alloy to behave in this manner is known as *quench sensitivity*.

Quench sensitivity. Quench sensitivity is probably due to a combination of two factors. Firstly, precipitation during the cooling process will arise if the cooling rate during the quench is sluggish. This rate will depend to a certain extent upon the nature of the quenching medium, but also of course upon the size and shape of the product being heat-treated, the rate of cooling being a function of the ratio of surface area to volume. The argument for a maximum quenching rate is not entirely one-sided, because both the degree of warpage or distortion that occurs during quenching and the magnitude of the residual stresses that develop in the products tend to increase with the rate of quenching.

The second important factor is the concentration of quenched-in vacancies, i.e. the degree of retention at the low temperature of the concentration of vacancies which existed in thermal equilibrium at the solution heat-treatment temperature. Zone formation in aluminium alloys at low ageing temperatures occurs at a rate of seven or eight *orders of magnitude* greater than that expected from the extrapolation of high-temperature diffusion data, because of the presence of these quenched-in vacancies. A slow quench will sharply reduce the concentration of these vacancies, and hence influence the subsequent rate of ageing. Therefore massive specimens would be expected to show a much lower hardening rate than small specimens, because of their inability to retain excess vacancies.

The range of temperature over which the quenching rate has its most critical influence on the mechanical properties of the aged material was first explored by Fink & Willey (1947) in the case of a high-strength Al-Zn-Mg alloy. Specimens were solution-treated and then held at various intermediate temperatures for various times before finally quenching to room temperature. The hardness of these specimens after subsequent ageing was then expressed in hardness contours on a graph whose axes were intermediate holding temperature and holding time, producing a series of 'C' curves (fig. 1.23) showing the tensile and yield strengths of the alloy expressed as percentages of strengths obtained by quenching without interruption.

Cooling rates that would be represented by time–temperature curves lying to the left of the 'C' curves of fig. 1.23 would permit development of virtually the maximum strengths of which the alloy is capable, whereas rates of cooling that would cut through the 'C' curves would not be sufficiently rapid to allow development of full strengths during subsequent ageing. Quench sensitivity is found to be dependent upon the constitution of the alloy, and this can at least be qualitatively understood. Thus the presence of *second phases* may nucleate precipitation during the quench. It is known, for example, that the presence of manganese or chromium enhances the sensitivity to rates of cooling in the quench in those aluminium-based alloys to which these elements are added to control grain growth by the formation of incoherent particles of an intermetallic phase. Coherent second phases are presumably much less likely to act as nucleants in this way. Again, a binding energy between vacancies and *solute atoms* would lead to a higher retained vacancy concentration, but the present lack of adequate experimental data preclude the drawing of firm conclusions on this point.

1.3.3   The control of precipitate distribution

Trace element effects

There appear to be at least four independent ways in which the addition of small quantities of solute to a given age-hardening alloy may have an influence upon the precipitation process.

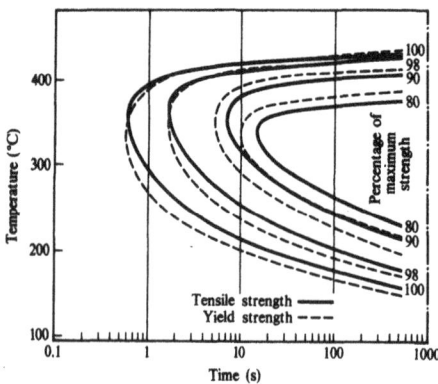

Fig. 1.23. Effect of time at constant temperature in the quenching range on tensile properties of an Al–Zn–Mg alloy. (After Fink & Willey, 1947.)

Vacancy interactions. There has been much recent work on trace element additions to aluminium-based alloys, and some of these additions have been shown to greatly affect the nature of the PFZ. For example, Polmear (1968) has shown that in the Al-Zn-Mg system, trace additions of silver stimulate the nucleation of the η'-phase, resulting in a finer, more uniformly distributed precipitate. On ageing above T_c, silver causes a marked reduction in the width of the PFZs, which implies that there is an interaction between the silver atoms and vacancies, so that the latter are retained within the matrix rather than migrating to the grain-boundary sink. Vacancy-solute interactions have also been observed in Al-Cu alloys, where trace additions of Cd, In or Sn have the effect of changing the kinetics of zone formation, and also have an influence upon the width of the PFZs.

Interfacial energy effects. By segregation to the interface between precipitate and matrix, an added trace element could change the interfacial energy and hence the precipitate morphology. Stewart & Martin (1970) have demonstrated this effect in the case of Al-Si alloys to which an addition of Cu was made. A simple binary alloy of Al and 0.5wt% Si, after solution-treatment and water quenching, will on ageing at, say, 230 °C precipitate Si as a mixture of submicron-sized plates and equiaxed particles. As the ageing temperature is increased over a small range, there is a marked increase in the particle size of the precipitate, so that coarse Si plates, several microns in diameter, are formed on ageing at 275 °C. The additions of 0.45wt% Cu to the alloy influences the morphology and growth kinetics of the Si precipitated: Equiaxed, rather than plate-like Si particles are formed, and their rate of growth upon increase in ageing time or temperature is markedly inhibited.

The effect of Cu appears to be principally on the growth rather than nucleation rate of the Si precipitate particles. This could be attributable to segregation of Cu either to the Si-particle/Al-matrix interface, or within the Al matrix adjacent to the particles, hence lowering the total interfacial energy.

Segregation to grain boundaries. There is evidence that certain solutes may segregate to the grain boundaries of a supersaturated solid solution and hence change the subsequent precipitation process during ageing. For example, when binary Cu-Be alloys are aged, after the initial formation of coherent zones of CuBe, the precipitation process proceeds by a discontinuous mechanism, in that cellular grain-boundary precipitates grow at the expense of the zones with associated grain-boundary migration. This combined precipitation and recrystallization process, which leads to

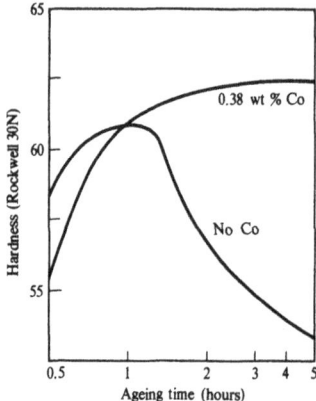

Fig. 1.24. The retardation of softening by the addition of 0.38wt% Co to an alloy Cu–2wt% Be when aged at 316 °C.

the formation of a relatively coarse dispersion of an intermetallic phase, is associated with a dramatic fall in hardness. However, commercial beryllium bronzes contain 0.2wt% of Ni or Co in addition to 1.9wt% Be. The former elements evidently segregate preferentially to the grain boundaries of the Cu for, upon ageing a ternary alloy, it is found that discontinuous precipitation is inhibited and a higher peak hardness may be obtained (fig. 1.24).

Change in free energy of precipitating phases. An added trace element may also be capable of bringing about a change in the free energy in such a way that a different phase is formed upon ageing. For example, when the ternary Al–2.5wt% Cu–1.5wt% Mg is aged, the first stage of ageing involves the formation of GP zones comprising ordered clusters of Cu and Mg atoms. These zones persist during the second stage of ageing in which precipitation of laths of the S′-precipitate occurs on dislocation lines generated in the quenching operation, or by cold working prior to ageing. (S′ is a distorted form of Al_2CuMg phase.) The addition of 0.5wt% Ag to the alloy greatly modifies the mechanism of ageing, in that homogeneous nucleation of the ternary T-phase takes place which predominates up to peak hardness, after which it is replaced by the S′- or S-phase. This type of effect could not arise through an interaction between Ag atoms and vacancies alone. It is considered that there is also an interaction between Ag atoms and Mg atoms, which modifies the nucleation processes of the particular precipitates involved.

Mechanical-thermal treatments (MTT)

When a combination of mechanical and thermal treatment is given to supersaturated alloys, advantageous changes in the distribution of precipitates may be brought about.

If an Al alloy is aged under conditions leading to the formation of pronounced PFZs due to vacancy depletion, when there is a large difference in yield stress between the PFZ and the precipitation-hardened grain interior, preferred deformation will occur within these weak zones, which can lead to a low macroscopic ductility in the alloy. If, after ageing, only limited deformation is applied and the alloy is re-aged, further precipitation occurs upon the dislocations introduced into the PFZ by the localized strain. Since the PFZ arose through vacancy depletion and not through solute depletion, the introduction of heterogeneous nucleation sites in the form of an array of dislocations leads to rapid precipitation.

More generally, the effect of mechanical-thermal treatment can either be to influence precipitation by working, or to combine work-hardening and age-hardening. Multiple mechanical-thermal treatment (MMTT) involves the repeated application of cycles of deformation and annealing or ageing, and has been applied to a number of materials, both ferrous and non-ferrous. We will defer a more detailed consideration of this subject until we have considered the deformation behaviour of crystals containing precipitates, confining our discussion here to a consideration of ageing processes in deformed metals.

When a supersaturated alloy is deformed and then aged, it is with the object of increasing the precipitate density by the introduction of prolific heterogeneous nucleation sites. The excess point defects introduced by the deformation may also accelerate the ageing process. In order for the operation to be successful it is obvious that the rate of precipitation must exceed the rate of recovery of the dislocation substructure. If the metal recrystallizes before precipitation commences, the deformation stage is of no value. The order of occurrence of precipitation and recrystallization in this situation has been considered by Köster (1971), who considered the temperature dependence of the start of both reactions in the following way.

The time for the start of precipitation (considering the formation of one phase only), t_p, may be written from (1.7) as

$$t_p = K_p \exp \left(\Delta G^* + U_D \right)/kT,$$

where K_p is a factor containing the driving force, an entropy term and geometric factors. The time for the start of recrystallization, t_r, may be written as

$$t_r = K_r \exp \left(\Delta G^r /kT \right),$$

where K_r is a factor containing the driving force for the reaction, an entropy term and geometric factors, and ΔG^r is the activation energy for

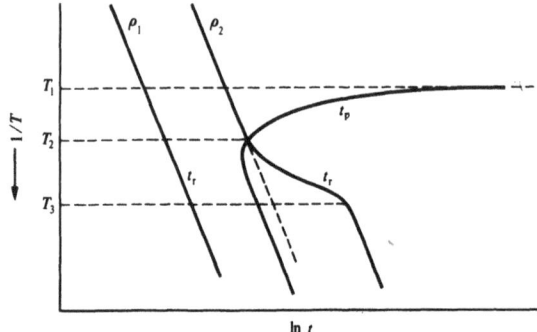

Fig. 1.25. Temperature dependence of the incubation time for precipitation, t_p and for recrystallization, t_r, as a function of dislocation density ρ. It is assumed here that the dislocations have little influence on the start of precipitation. With dislocation density ρ_1, recrystallization is always complete before precipitation. For dislocation density ρ_2 ($\rho_2 < \rho_1$): with $T > T_1$, the single phase is stable; for $T_2 > T > T_3$ discontinuous recrystallization occurs, but is slowed by the pinning effect of the precipitates; and for $T_3 > T$ continuous recrystallization only occurs. (Schematic diagram after Köster, 1970.)

the formation of a recrystallization front. ΔG^r has the order of magnitude of the activation energy for self-diffusion, but decreases as the dislocation density increases, and increases as a result of segregation of solute atoms at dislocations.

Curves of t_p and t_r as a function of temperature and increasing annealing time are shown in fig. 1.25. Three temperature ranges in the recrystallization behaviour can be distinguished on the basis of this diagram.

(i) When the temperature is above T_1 (the solvus temperature) no precipitation occurs, and recrystallization is influenced only by segregation.

(ii) When the temperature is between T_1 and T_2 precipitation proceeds in the recrystallized structure after completion of recrystallization. Again, recrystallization is influenced only by segregation of solute.

(iii) At temperatures below T_2, precipitation proceeds in the deformed structure, which is the situation sought in MTT. Recrystallization itself will be influenced by the presence of the particles formed, and this is indicated in fig. 1.25 by a deviation of the curve t_r from linearity towards longer recrystallization times. In this temperature range the precipitating particles influence both the rearrangement of dislocations to form recrystallization fronts and their migration.

For successful MTT therefore, it is important that temperature T_2 should be as high as possible. This implies that the curve of t_p is as far as possible to the left, and that t_r is as far as possible to the right in fig. 1.25. For the first requirement, U_D and ΔG^* should be minimized, so that systems involving a sluggish diffusion of substitutional solute atoms are unlikely to lend themselves to MTT, whereas systems involving interstitial elements, such as carbon in steel, are more likely to respond.

We will discuss the extent to which structures of high strength may be achieved by these means at the end of chapter 2.

1.4   Quantitative metallography

In order to consider in a quantitative way the relationship between the mechanical properties and the microstructure of a dispersed-phase alloy, the distribution of the precipitate must be defined. This is a formidable problem in view of the wide variations in size and shape distribution which are found in precipitated systems. We will, therefore, initially consider a simplified idealized situation.

1.4.1   A dispersion of uniform spherical particles of radius *r*

The volume fraction, *f*, of the dispersed phase is defined as the ratio of the volume occupied by the particles to the total volume. In a perfectly randomly dispersed microstructure of particles in a matrix, *f* is also equal to the fractional area occupied by the particles observed on a random polished cross-section, and is also equal to the fractional length occupied by the particles on a random straight line taken through the structure.

If N_V is the number of spherical particles per unit volume, then the definition of *f* yields:

$$f = \tfrac{4}{3}\pi r^3 N_V. \tag{1.19}$$

N_V is not, however, a parameter which is readily measured by metallographic means. A polished cross-section of the structure will reveal the particles as a series of circular areas of radius between 0 and *r*. Then N_S, the number of particles per unit surface area may easily be obtained. Let us consider the relationship between N_V and N_S.

Consider a unit cube of the structure, and take a section parallel to one of the cube faces. Then the number of intersections with particles (N_S) will be given by

$$N_S = N_V \, p,$$

where *p* is the probability that the plane will intersect a single particle placed randomly within the cube. Since, of the various positions of the cross-sectional plane, only those positions existing over the length 2*r* would lead to the plane intersecting the single sphere, the probability of intersection is equal to 2*r*, thus

$$N_S = 2N_V \, r. \tag{1.20}$$

Applying the equality of area and volume fractions,

$$f = \tfrac{4}{3}\pi r^3 N_V$$

$$= N_S \bar{a},$$

where $\bar{a}$ is the average area of particle intersected by a random plane of cross-section, i.e.

$$f = 2N_V \, r \, \bar{a}$$

or

$$\bar{a} = \tfrac{2}{3}\pi r^2. \tag{1.21}$$

Although this provides a means of measuring the true particle size in an opaque sample studied in a random planar section, the analysis of the data is tedious, and further simplification may be obtained.

Lineal analysis. When random lines are passed through the sample parallel to a cube edge, the number of spheres intersected by the line (N_L) is given by

$$N_L = N_V \, p',$$

where p' is the probability of the line hitting a single randomly placed sphere within the cube. Since possible positions of the line occupy unit area, and the possible positions for which it will pass through the sphere occupy an area of πr^2, $p' = \pi r^2$. Therefore

$$N_L = \pi r^2 \, N_V.$$

From (1.19)

$$f = \tfrac{4}{3}\pi r^3 N_V = N_L \, \bar{l}$$

and therefore

$$\bar{l} = \tfrac{4}{3}r, \tag{1.22}$$

where $\bar{l}$ is the mean lineal traverse length of the random line over the particles. Combining equations (1.20) and (1.22), the particle radius is obtained

$$r = 2N_L /\pi N_S, \tag{1.23}$$

so that one may obtain a value for r simply by counting intercepts – lengths or area need not be measured.

Again, from (1.20) and (1.22), eliminating r, we have

$$N_V = \tfrac{1}{4}\pi \, N_S^2 /N_L, \tag{1.24}$$

so that, by substitution of (1.23) and (1.24) into the equation for f we obtain

$$f = 8 N_L^2 / \pi N_S. \tag{1.25}$$

Equations (1.23) and (1.25) also yield

$$f = \tfrac{4}{3} r N_L, \tag{1.26}$$

which provide means of measuring the volume fraction of a phase dispersed as uniform spheres. In practice these relationships may be used to give approximate measurements of the dispersion parameters provided that the range of particle sizes actually present is small.

Particle spacing parameters. The basic characteristics of a dispersed-phase structure can be described in terms of f and r. The interparticle spacing is a function of these parameters, and it is calculated by their substitution into one of several possible formulae, which have been reviewed by Corti, Cotterill & Fitzpatrick (1974).

The mean free path (MFP)
The MFP is defined as the average distance between the surfaces of particles along any straight line in the structure. In alloys with random particle distributions the value of the MFP is independent of the direction of the chosen line, and is given by

$$\mathrm{MFP} = (1 - f)/N_L. \tag{1.27}$$

Combining (1.26) and (1.27), we can write

$$\mathrm{MFP} = \tfrac{4}{3} r (1 - f)/f. \tag{1.28}$$

Mean planar separation of particles
In chapters 2 to 4 we shall attempt to establish the relationship between the microstructure and the mechanical properties of dispersed-phase alloys. Many different spacings can be calculated. The appropriate spacing and method of averaging for a particular problem depends on the nature of the problem.

The average centre-to-centre spacings, λ, between a particle and its *nearest neighbour* intersected by a random plane through the array is

$$\lambda_S = 0.5 N_S^{-\frac{1}{2}}. \tag{1.29}$$

This nearest-neighbour spacing is not, however, always appropriate. Thus, in calculating the interaction between a gliding dislocation with a random array of obstacles in its slip plane, the average distance from a particle to its nearest two, three or four neighbours (instead of to its one nearest neighbour) is the correct estimate. Kocks (1966) and also Foreman &

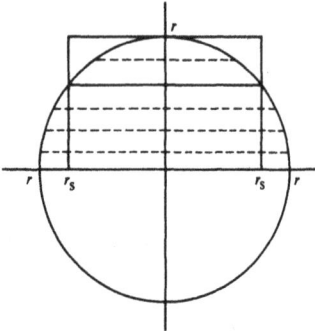

Fig. 1.26. Definition of mean planar particle radius r_S of cross-section of particle of radius r.

Makin (1966) have considered this problem, and in general the more nearest neighbours that are taken into account the longer will be the resulting evaluated separation (λ_S^*). They obtain

$$\lambda_S^* = 1.8 \, (N_S)^{-\frac{1}{2}} = 1.25r \, (2\pi/3f)^{\frac{1}{2}}. \tag{1.30}$$

If the size of the particles is negligible in comparison with their centre-to-centre separation, i.e. if $\lambda \gg r$, then (1.30) is an appropriate description of the microstructure. If this is not the case, then one must consider the surface-to-surface mean planar separation of the particles, λ_S^{**}, which reduces λ_S^* by an amount $2r_S$, where r_S is the mean radius of intersection of a random plane with a spherical particle of radius r.

The value of r_S can be obtained by considering a particle of radius r intersected by a series of parallel planes (see fig. 1.26). The mean radius of intersection, r_S, is obtained by equating the volume of the hemisphere of radius r to the volume of the cylinder radius r_S and height r

$$\tfrac{2}{3}\pi r^3 = \pi r_S^2 r$$

i.e.
$$r_S = \sqrt{\tfrac{2}{3}} r. \tag{1.31}$$

The mean planar surface-to-surface separation thus becomes

$$\lambda_S^{**} = \lambda_S^* - 2\sqrt{\tfrac{2}{3}} r. \tag{1.32}$$

1.4.2 Dispersions of uniform plate- or rod-like particles

Fullman (1953) has extended the particle size analysis discussed above to the cases of structures consisting (i) of dispersions of circular plates of uniform radius and thickness and (ii) of dispersion of uniform rods. We will not pursue this refinement here, since the simplifying assumption of equiaxed particles is usually made in structure/property relationships.

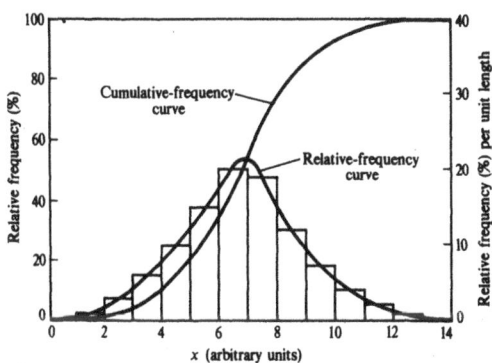

Fig. 1.27. Relative-frequency and cumulative-frequency curves.

1.4.3   The size distribution of spherical particles

Microstructure in real dispersed-phase alloys do not contain dispersions of spheres of uniform size, and metallographic analysis will reveal a *distribution* of sizes in each of the several phases that may be present (see §1.1). This subject has been reviewed in some detail by Exner (1972), who has discussed how experimental data on particle size may be processed by representing it graphically or by analytical distribution functions.

The most convenient graphical representation is to plot a histogram (fig. 1.27) illustrating the relative frequency of occurrence of the classes of particle size observed in a given phase. From the histogram a cumulative-frequency distribution curve can be plotted (fig. 1.27) over the upper limits of each class. The continuous cumulative-frequency distribution curve runs from x_{min} to x_{max} (where x is the size parameter measured).

For comparing size distributions of similar shape and for the presentation of results it is desirable to know the mathematical function of the size distribution. In general the log–normal distribution, i.e. the Gaussian distribution with a logarithmic abscissa, is used for describing size distributions of grains and particles. If N is the frequency of observation of particle size d and the total number of observations is ΣN, then the arithmetic mean particle size is given by

$$M_a = \Sigma Nd/\Sigma N. \tag{1.33}$$

The standard deviation (σ_a) is a measure of the dispersion of the observations and is given by

$$\sigma_a = (\Sigma [N(d - M_a)^2]/\Sigma N)^{\frac{1}{2}}. \tag{1.34}$$

In the case of log-normal distribution, it is usual to calculate the geometric mean, given by

$$\log M_g = \Sigma (N \log d)/\Sigma N, \tag{1.35}$$

The standard geometric deviation is given by

$$\log \sigma_g = (\Sigma \left[(\log d - \log M_g)^2 \right] / \Sigma N)^{\frac{1}{2}}. \tag{1.36}$$

The equation to the curve describing the log–normal size distribution (fig. 1.26) is then

$$N = [\Sigma N / (2\pi)^{\frac{1}{2}} \log \sigma_g] \exp \left[-(\log d - \log M_g)^2 / 2 (\log \sigma_g)^2 \right]. \tag{1.37}$$

Calculation of M and σ can be laborious, but it is simplified through the use of a graphical solution through the use of logarithmic probability paper. When the cumulative-frequency curve is plotted in this way a linear relationship is obtained. Fig. 1.28 illustrates a range of experimental data plotted in this way, which demonstrates the log–normal distribution of particle sizes in a number of precipitated alloy systems as well as the log–normal distribution of grain sizes observed in a specimen of α-brass.

M_g is given by the size corresponding to 50% frequency value on such a plot, and σ_g may be obtained from the ratio

$$\sigma_g = \frac{84.13\% \text{ value}}{50\% \text{ value}} = \frac{50\% \text{ value}}{15.87\% \text{ value}}.$$

Exner (1972) concludes that, in spite of the fact that no satisfactory theoretical explanation has been advanced for its occurrence in microstructures, the log–normal distribution should be applied whenever it expresses the experimental results within the limits of experimental error.

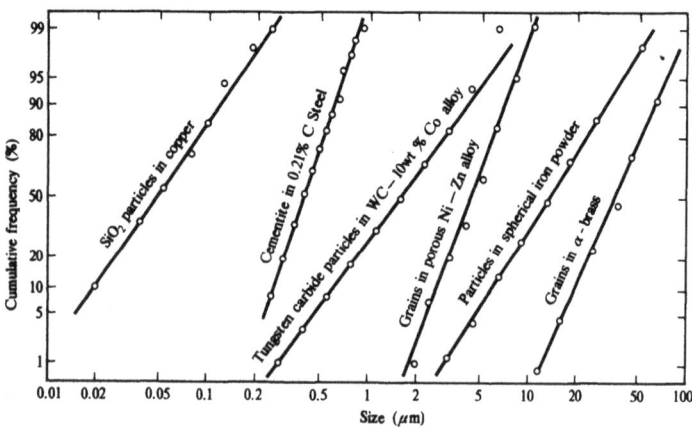

Fig. 1.28. Size distribution of grains and particles of various alloys in the log-normal probability graph. (After Exner, 1972.)

There are well-known relations between geometric and arithmetic means and standard deviations. Thus, if $\bar{d}_g$ and $\bar{d}_a$ are the geometric and arithmetic mean particle sizes, then

$$\ln \bar{d}_g = \ln \bar{d}_a - 0.5 \, (\sigma_a/\bar{d}_a)^2, \tag{1.38a}$$

$$\ln^2 \sigma_g = (\sigma_a/\bar{d}_a)^2. \tag{1.38b}$$

Once the particle-size distribution has been characterized, it is possible to obtain a value for the 'average particle spacing'. If (1.30) is written in terms of particle diameter, $d = 2r$, it becomes

$$\lambda_S^* = 1.2 \, (\pi/6f)^2 \tag{1.39a}$$

and the average planar spacing, when a distribution of particle sizes is present, may be written (Ashby & Ebeling, 1966)

$$\lambda_S^* = 1.2 \, [(\pi/6f) \, (1 + \sigma_a^2/\bar{d}_a^2)]^{\frac{1}{2}} \, \bar{d}_a. \tag{1.39b}$$

Particle-size distributions during precipitate coarsening. If dispersed-phase alloys are held at elevated temperatures after precipitation is complete, the precipitate particles progressively coarsen, the larger particles growing at the expense of the smaller ones. The average interparticle spacing thus increases with time. The driving force for this process, known as *Ostwald ripening*, is the tendency to reduce the total area of internal interfaces, and will be greater the higher the interfacial energy.

The kinetics of the coarsening process has been treated theoretically by Wagner (1961) and others, and the rate-controlling mechanism can be either the solution reaction at the particle/matrix interface, or the diffusion of the solute through the matrix. Wagner has shown that the 'steady-state' distribution of particle sizes, $f(r, t)$ to which any initial distribution evolves, is dependent upon which of the two processes is rate-controlling.

In the case of diffusion control

$$f(r, t) = f'(t) \, \rho^2 \, h(\rho), \tag{1.40}$$

where $\rho = r/\bar{r}$ and $h(\rho)$ is given by

$$h(\rho) = \left(\frac{3}{3+\rho}\right)^{\frac{7}{3}} \left(\frac{\frac{3}{2}}{\frac{3}{2} - \rho}\right)^{\frac{11}{3}} \exp\left(\frac{-\rho}{\frac{3}{2} - \rho}\right) \quad \text{for } \rho < \frac{3}{2}$$

$$h(\rho) = 0 \quad \text{for } \rho \geqslant \frac{3}{2}$$

where $\bar{r}$ is the arithmetic mean value of particle radius (r), and $f'(t)$ is a function of time only. The function $\rho^2 h(\rho)$ has the following characteristic features:

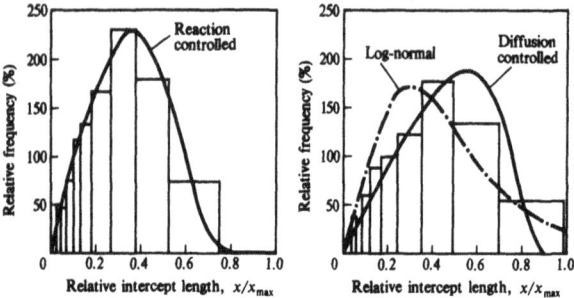

Fig. 1.29. Comparison of theoretical (shown by smooth lines) and experimental (shown by histogram) intercept distributions of VC grains in an Ni matrix (sample annealed for 16 hours at 1500 °C). Agreement is found for reaction control while considerable deviation occurs between the experimental frequencies and the curves predicted for diffusion control as well as for the log-normal grain-size distribution often observed experimentally in two-phase microstructures. (After Exner, Santa Marta & Petzow, 1971.)

(i) a sharp cut-off at $\rho = \frac{3}{2}$;
(ii) a maximum at $\rho = \frac{3}{2}$;
(iii)

$$\int_0^\infty \rho^2 h(\rho)\, d\rho = \frac{9}{4}.$$

In the case of reaction control,

$$f(r,\ t) = f''(t)\,\rho h'(\rho), \tag{1.41}$$

where $h'(\rho)$ is given by

$$h'(\rho) = \left(\frac{2}{2-\rho}\right)^5 \exp\left(\frac{3\rho}{\rho-2}\right) \quad \text{for } \rho < 2,$$

$$h'(\rho) = 0 \quad \text{for } \rho \geqslant 2.$$

The difference between the particle-size distributions given by (1.40) and (1.41) has been used by Exner, Santa Marta & Petzow (1971) to identify the mechanism of growth of particles of vanadium carbide in nickel during liquid-phase sintering (i.e. sintering at temperatures above melting-point of the nickel matrix). This resulted in a regular structure of nearly spherical VC grains in a Ni matrix. The particle-size distribution was measured metallographically and compared with that predicted by (1.40) and (1.41) and also with a log–normal distribution. Their results, shown in fig. 1.29, indicate that the observed size distribution of their particles of

VC was consistent only with coarsening in which the rate is controlled by *reaction*, i.e. the solution and precipitation of VC in the liquid metal, and not by the rate of diffusion of atoms within the matrix.

Having discussed the *structure* of precipitation-hardened alloys, we are now in a position to consider its relationship with mechanical properties. In the following chapter we will therefore examine the yield and work-hardening behaviour of particle-hardened structures.

2 Yield and work-hardening in the absence of recovery

2.1 The yield stress of two-phase alloys

The stress required to initiate plastic flow can be calculated with some confidence: it is the stress required to send a dislocation large distances through the array of precipitates. This stress is a threshold stress for plastic flow, and the theoretical treatment of second-phase particle strengthening is based on the following model. The glide dislocation bows out between the particles as shown in fig. 2.1 and, when the included angle ϕ between the two arms of the dislocation reaches a certain critical value, the dislocation breaks away from the obstacle. We will refer to this value of ϕ as the *breaking angle*, and at this critical point the obstacle strength, F, is related to the dislocation line tension, T, by

$$F = 2\,T \cos \tfrac{1}{2} \phi \tag{2.1}$$

When the breaking angle $\phi = 0$, the particle behaves as an impenetrable obstacle, while for values of $\phi > 0$, the particle can be sheared by the glide dislocation, with the required shearing force equal to F. We will consider these two types of interaction separately.

2.1.1 Particle shearing

The shear stress needed to cause the dislocation to break away from the particle may be related to F as follows. The applied shear stress, τ, causes the dislocation of Burgers vector b to bow into a loop of radius of curvature R, where

$$\tau\,b = T/R. \tag{2.2}$$

From fig. 2.1 it can be seen that

$$2\,R \sin \theta = \lambda, \tag{2.3}$$

so

$$\tau\,b = \frac{2\,T \sin \theta}{\lambda}.$$

But it can also be seen that $\theta = 90° - \tfrac{1}{2}\phi$, hence from (2.1) we have

$$\tau\,b = F/\lambda,$$

i.e.

$$\tau\,b\,\lambda = F. \tag{2.4}$$

One of the vital terms in this fundamental strengthening expression is the

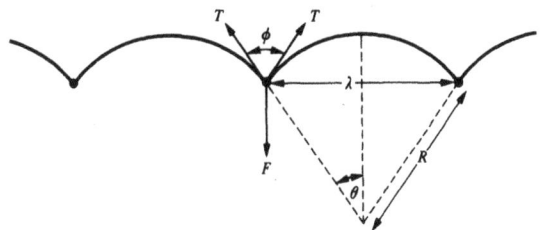

Fig. 2.1. Dislocation held up at obstacles. The bowing angle ϕ is defined as shown.

effective interparticle spacing. In the case of particles that can be sheared by the glide dislocation, the value of this parameter is given approximately by the *Friedel relationship.*

Friedel assumes that during the yielding process the dislocation takes up a *steady-state* configuration, namely that each time a dislocation breaks through one obstacle (B) (fig. 2.2) it meets one, and only one, other obstacle (B') as it bows out to the configuration compatible with the stress applied (fig. 2.1). For many alloys of practical interest (e.g. aluminium-based alloys), the concentration of precipitate particles is sufficiently small to enable this 'point obstacle' theory to be used.

Each time a particle is sheared, therefore, an area A of the slip plane (shown shaded in fig. 2.2) is swept out by the dislocation. On average, the value of A is thus inversely proportional to the number of particles per unit area of slip plane. If the average spacing of particles in the slip plane is l, and if N_S is the number of particles per unit area of slip plane, then for a regular square array of particles

$$l = N_S^{-\frac{1}{2}} \tag{2.5}$$

and

$$A = l^2.$$

From fig. 2.2, the area swept out per particle shearing is given approximately by $h\lambda$, so

$$l^2 = h\lambda. \tag{2.6}$$

Also, from the property of the circle (fig. 2.2),

$$\lambda^2 = 2hR, \quad \text{for } h \ll R. \tag{2.7}$$

Eliminating h from (2.6) and (2.7), we obtain

$$\lambda^3 = 2l^2 R = l^2 \lambda/\cos\tfrac{1}{2}\phi. \tag{2.8}$$

The spacing of the obstacles thus depends upon the value of the breaking angle, ϕ, i.e. on F (2.1), so that the stronger the obstacles the smaller the value of λ.

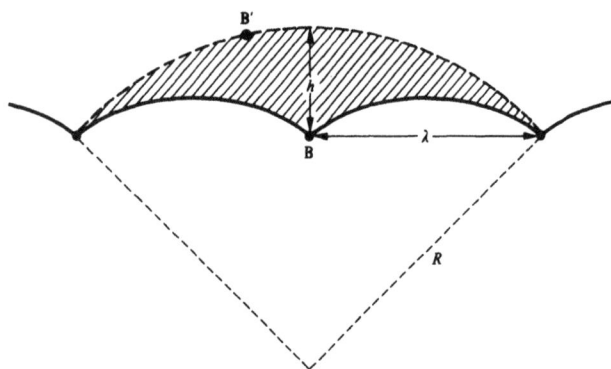

Fig. 2.2. The Friedel process for dislocations interacting with point obstacles.

Equation (2.8) may be written

$$(\lambda/l)^2 = \sec \tfrac{1}{2}\phi = 2T/F,$$

and substituting for λ in (2.4) we obtain a value for the yield stress,

$$\tau = F^{\frac{3}{2}}/b \, l \, (2T)^{\frac{1}{2}}. \tag{2.9}$$

Substitution of (2.1) for F gives

$$\tau = 2T (\cos \tfrac{1}{2}\phi)^{\frac{3}{2}}/bl. \tag{2.10}$$

This equation has been derived assuming small breaking angles and a regular square array. Computer calculations by Foreman & Makin (1967) for a *random* array of obstacles showed that this equation is a good approximation, except for strong obstacles where $\phi \to 0$. These workers therefore suggested the following empirical equation, which applies over the whole range of obstacle strengths,

$$\tau = 2T (0.8 + \tfrac{1}{5} \phi/\pi) \cos (\tfrac{1}{2}\phi)^{\frac{3}{2}}/bl. \tag{2.11}$$

In order, therefore, to obtain a theoretical estimate of the yield stress in the case of particles that are cut by glide dislocations ($\phi > 0$), the force, F, required to cut the particle (2.9) *or* the value of the breaking angle ϕ (2.10) must be calculated. There are a number of possible sources for this shearing force F.

(i) *Coherency hardening* arises from the elastic coherency stresses surrounding a particle that does not fit the matrix exactly.

(ii) *Surface or chemical hardening* arises from the energy required to create an additional particle/matrix interface when the particle is sheared by the dislocation.

(iii) *Order hardening* is due to the additional work required to create an antiphase boundary in the case of dislocations passing through precipitates which have an *ordered* lattice.

(iv) *Stacking-fault hardening* occurs when there is a difference between the stacking-fault energy of the particle and that of the matrix when these are either both face-centred cubic (fcc) or both hexagonal close-packed (hcp) in structure.

(v) *Modulus hardening* arises from differences between the elastic moduli of matrix and particle.

We will consider these mechanisms in turn, and examine the evidence from experiments on real materials which support the theoretical predictions.

Coherency hardening

We will use (2.9) to derive a value for the yield stress, τ, when F arises from coherency strains associated with the precipitates. From (2.5), $l = N_S^{-\frac{1}{2}}$, and if r is the particle radius and f the volume fraction of the precipitate, we can write, (see 1.20),

$$\frac{N_S}{2r} = \frac{3f}{4\pi r^3},$$

i.e.

$$N_S = \frac{3f}{2\pi r^2} = \frac{1}{l^2}. \tag{2.12}$$

For a completely spherical inclusion of radius r containing material of atomic volume $(1 + \delta)^3$, where the atomic volume of the matrix is unity, the stress in the precipitate is a uniform pressure

$$p = 3K(\delta - e),$$

where

$$e = \frac{e K \delta}{3K + 2E/(1 + \nu)}.$$

K is the bulk modulus of the precipitate, and E and ν are Young's modulus and Poisson's ratio of the matrix. The strain in the matrix is a shear without a dilation. The shear strain at a distance R from the precipitate is given by er^3/R^3. The maximum strain and stress occur at the particle interface, i.e. at $R = r$, and for edge dislocations the appropriate stress is approximately Ge. Thus, as a dislocation approaches a particle it will experience a force per unit length of Geb. This force acts effectively over a length of dislocation equal to the particle diameter, hence

$$F \approx 2\,G\,e\,r\,b,$$

or
$$F = k\,G\,e\,r\,b, \tag{2.13}$$

where k is a numerical constant which a more accurate treatment gives as between 3 and 4. Substituting (2.13) and (2.12) in (2.9), and putting $T = \frac{1}{2}G\,b^2$, we find

$$\tau = k^{\frac{3}{2}}\,G\,e^{\frac{3}{2}}\,(r/b)^{\frac{1}{2}}\,(3/2\pi)^{\frac{1}{2}}\,f^{\frac{1}{2}}. \tag{2.14}$$

The yield stress due to the presence of the particles is thus predicted to be proportional to $f^{\frac{1}{2}}$ and also to $r^{\frac{1}{2}}$.

In seeking experimental support for this model, there is always difficulty in finding an alloy system that exhibits only one type of hardening. The question of how to combine more than one strengthening effect has received little attention. Witt & Gerold (1969) have measured the critical resolved shear stress of Cu–Co alloys containing coherent precipitates of fcc Co particles, and their data were consistent with the predictions of (2.14). By varying the Co content and the ageing treatment, both f and r were varied, and as seen in fig. 2.3 their data fitted a straight line up to $(r/b)^{\frac{1}{2}} = 5.3$ with r taken as 7.2 nm. Taking a misfit parameter value $e = 0.016$, as expected from structural data, the experimental results yield a slope about one-half of the value calculated from (2.14) which suggests some shortcoming in the present theory.

Surface, or chemical hardening

Harkness & Hren (1970) have considered the contribution to the yield stress ($\Delta\tau$) of the increase in precipitate/matrix interfacial area when a dislocation shears a particle. They considered the energy balance between

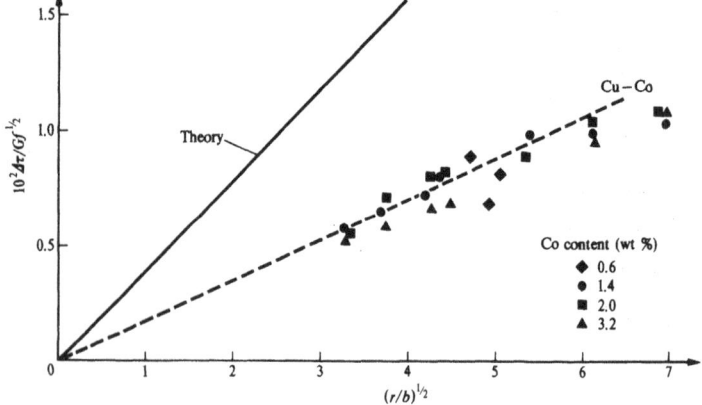

Fig. 2.3. Yield stress data for Cu–Co alloys. (After Witt & Gerold, 1969).

the work expended during yielding in the creation of a new surface area,
i.e.

$$F b = \sigma_S \Delta A \qquad (2.15)$$

where F is the average force acting over the incremental area, ΔA, and
where σ_S is the energy per unit area. Since F corresponds to the obstacle
strength, we can write, from (2.4),

$$F b = \Delta \tau\, b^2\, \lambda\, \langle d \rangle / \langle d_{max} \rangle. \qquad (2.16)$$

Since the increase in surface area of the interface will depend upon the
distance of the slip plane above the 'equator' of the particle upon which
the dislocation passes, the ratio of average chord length to maximum
chord length, $\langle d \rangle / \langle d_{max} \rangle$, is equal to the ratio of the average to maximum
area created during slip.

Harkness & Hren measured the increases in the critical resolved shear
stress ($\Delta \tau$) of Al–Zn single crystals at 77 K as a function of the volume
fraction and average size of the GP zone population. From a knowledge
of the geometry of the dispersed phase, they calculated the value of ΔA
(2.15), and fig. 2.4 illustrates their results equating the work expended
during slip to the corresponding area of precipitate/matrix interface
created. This plot includes results from zone volume fractions ranging
from 1.45% to over 30%, and for average zone diameters ranging from
0.8 to 9 nm. Though some scatter exists, the result is a linear graph whose
slope corresponds to the specific energy of surface creation, σ_S, and gives
a value for the energy per square metre of 0.32 J, which is within a factor
of two agreement with experimental measurement of this parameter.

A simple estimate of the yield-stress increment arising from this chemi-

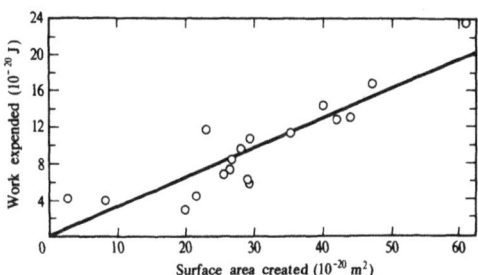

Fig. 2.4. The relationship between the work expended during slip in two-
phase Al–Zn crystals at 77 K and the area of precipitate/matrix interface
created. (After Harkness & Hren, 1970.)

cal hardening may be made by considering the force upon a screw dislocation as it passes through a coherent precipitate (fig. 2.5). When the dislocation advances by δx, the surface area increases by $2b\delta x$. Applying the principle of virtual work, therefore,

$$F = 2\sigma_S b, \tag{2.17}$$

and from (2.9), (2.12) and (2.17), putting $T = \frac{1}{2}Gb^2$, we obtain

$$\tau = 2(3/\pi)^{\frac{1}{2}} G (\sigma_S/Gb)^{\frac{3}{2}} (b/r) f^{\frac{1}{2}}. \tag{2.18}$$

The situation for an edge dislocation is similar. Equation (2.18) shows that if ageing occurs at a given volume fraction, then the maximum strength occurs at the minimum particle size. Brown & Ham (1971) suggest that zones in Cu-Be alloys cause strengthening through chemical hardening, but no unambiguous data are available.

Order hardening

When a dislocation passes through an ordered precipitate, an antiphase boundary is formed, of surface energy σ_{apb} per unit area. An estimate of the maximum force exerted by a spherical particle may be obtained by assuming the dislocation lies along a diameter, in which case

$$F = 2r\,\sigma_{apb}. \tag{2.19}$$

From (2.9), (2.12) and (2.19) we have

$$\tau = \frac{\sigma_{apb}}{2b} \left[\frac{3\,\sigma_{apb}\,f\,r}{\pi T} \right]^{\frac{1}{2}}. \tag{2.20}$$

Equation (2.20) explains the increase of τ on ageing in terms of the increase in r for a given volume fraction, and indicates why order hardening is thought to be a very important hardening mechanism in nickel alloys in which large volume fractions of the ordered γ'-phase (Ni_3Al, Ni_3Ti or Ni_3Al, Ti) may be present.

A complication which commonly arises in alloys containing ordered

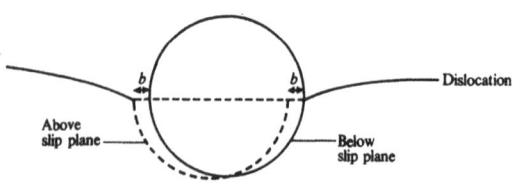

Fig. 2.5. A precipitate particle being sheared by a screw dislocation.

precipitates is that the dislocations travel in pairs, the second dislocation removing the disorder (the antiphase boundary) created by the first (as illustrated in fig. 2.6). Brown & Ham (1971) have calculated that when the first dislocation meets the Friedel condition, the second dislocation is pulled forward by the antiphase domain boundary remaining in the particles which it intersects. As τ is increased from zero, the first dislocation bends forward more, while the second dislocation straightens out, and forward stress on the first dislocation is substantially increased by the interaction of the second dislocation with disordered particles. Depending on the particle size, r, Brown & Ham have calculated two expressions for τ for paired dislocation movements

$$\tau = \left(\frac{\sigma_{apb}}{2b}\right)\left[\left(\frac{4\sigma_{apb}\,rf}{\pi\,T}\right)^{\frac{1}{2}} - f\right], \quad \text{for } \pi Tf/4\sigma_{apb} < r < T/\sigma_{apb},$$

(2.21)

$$\tau = \left(\frac{\sigma_{apb}}{2b}\right)\left[\left(\frac{4f}{\pi}\right)^{\frac{1}{2}} - f\right], \quad \text{for } r > T/\sigma_{apb}.$$ (2.22)

Chaturvedi, Lloyd & Chung (1976) have recently studied the yielding behaviour of γ'-phase precipitation-strengthened Co–Ni–Cr superalloys containing various volume fractions of the ordered Ni_3Ti-phase. When these γ'-particles were smaller than 2.8 nm in diameter they were sheared by the glide dislocations. It has been observed that the contribution of modulus strengthening by γ'-phase in Ni-based alloys is very small, and can be ignored, as is also the case with chemical hardening. Similarly, stacking-fault strengthening (see below) is not expected to contribute significantly to the total strength of the alloy, so only coherency and order-hardening mechanisms were considered in the interpretation of their experimental data.

Fig. 2.7 shows plots of the observed values and of the calculated values

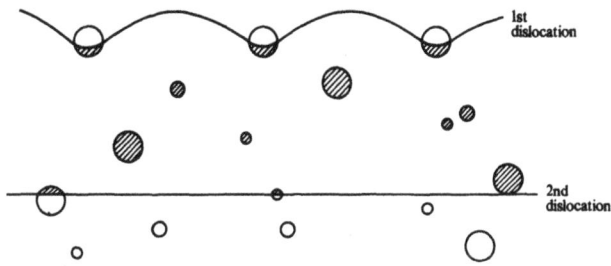

Fig. 2.6. Movement of paired dislocations through a crystal containing precipitates of an ordered phase: shaded areas show the existence of anti-phase boundaries within the particles.

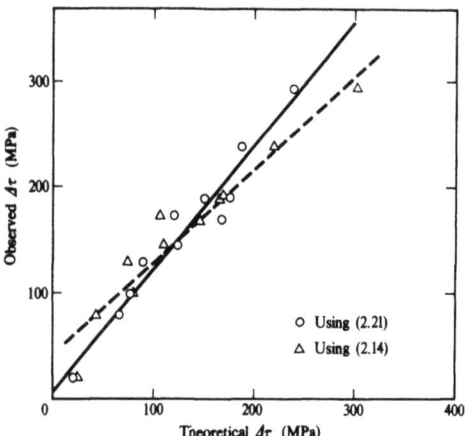

Fig. 2.7. Plots of observed versus theoretical values of $\Delta\tau$ during the early stages of ageing of a Co–Ni–Cr superalloy. (After Chaturvedi, Lloyd & Chung, 1976.)

of the shear-stress increment $\Delta\tau$ arising from coherency hardening (2.14), and from order hardening (2.21). It can be seen that the experimental data is in reasonable agreement with *each* model. In order to decide which mechanism was operative in controlling the strength, plots were made of $\Delta\tau$ versus $r^{\frac{1}{2}}$. If coherency hardening were operative, then (2.14) predicts that the graph should be linear and pass through the origin, whereas according to (2.21) the straight line should not pass through the origin. In fact the data, when so plotted, did not pass linearly through the origin (fig. 2.8), and it was concluded that when, in this alloy system, the precipitate particles are sheared by the glide dislocations, the yield stress is governed by the order-strengthening mechanism.

Stacking-fault hardening

The energy of an extended dislocation is decreased if the energy of the stacking-fault formed between its component partial dislocations is reduced. Hirsch & Kelly (1965) were the first to suggest that there should be a strong interaction between dislocations and particles in which the stacking-fault energy σ_{sfp} is much lower than that of the matrix, σ_{sfm}.

When a long straight dislocation dissociates into partials with an equilibrium ribbon width w_{m}, moves into a large particle of lower stacking-fault energy and splits further into a ribbon of width $w_{\text{p}} > w_{\text{m}}$ (fig. 2.9), the reduction in energy per unit length is

$$\Delta E = K(\theta) \ln(w_{\text{p}}/w_{\text{m}}).$$

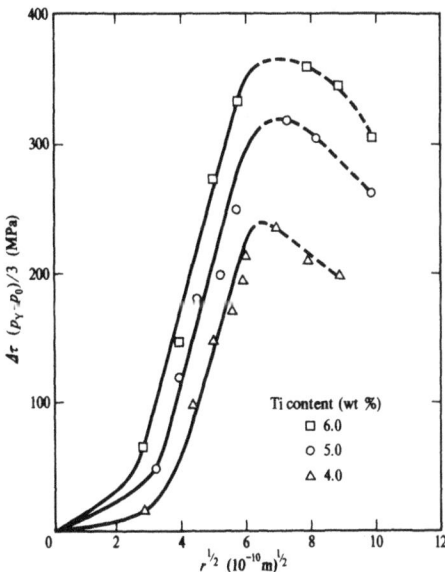

Fig. 2.8. Variation in $\Delta\tau$ with $r^{\frac{1}{2}}$ in Co–Ni–Cr superalloy. (After Chaturvedi, Lloyd & Chung, 1976.)

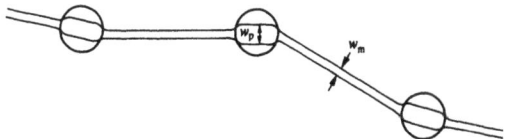

Fig. 2.9. Configuration of dislocation dissociated into partials of equilibrium ribbon width w_m in the matrix and width w_p within the particles.

where $K(\theta)$, in a close packed metal, is a factor approximately equal to $G\,b_p^2/4\pi$, where b_p is the length of the Burgers vector of either partial dislocation. $K(\theta)$ depends upon θ (the angle between the dislocation line and the Burgers vector) in such a way that it is greater for an edge dislocation than it is for a screw (by a factor of 2.33 when Poisson's ratio is 0.33).

The dislocation must be dragged out of the particles with a local stress τ' given approximately by

$$\tau' \approx \Delta\sigma/b, \tag{2.23}$$

where $\Delta\sigma = \sigma_{sfm} - \sigma_{sfp}$. This stress is opposite in sign but of comparable magnitude with that for order strengthening. This mechanism leads to a variation of τ with r given by

$$\tau = \left(\frac{8}{\pi}\right)^{\frac{1}{2}} G \left(\frac{\Delta\sigma}{Gb}\right)^{\frac{3}{2}} \left(\frac{r}{b}\right)^{\frac{1}{2}} f^{\frac{1}{2}} I_m, \tag{2.24}$$

where I_m is a complex function of r and the stacking-fault energy of the precipitate. Order-of-magnitude estimates of τ using this equation are in reasonable agreement with experiment. In systems such as Al–4wt% Ag the dislocations may be attracted to the GP zones (fig. 2.9) because the stacking-fault energy of Ag (approximately 20 mJ m^{-2}) is less than that for Al (approximately 200 mJ m^{-2}).

Modulus hardening
Since the energy of a dislocation is a function of the shear modulus of the lattice in which the strain field of the dislocation exists, a change in energy, and hence a force, will be associated with a dislocation interacting with a particle whose shear modulus differs from that of the matrix. An extreme case is the interaction with a void, which is a region in which the elastic energy is reduced to zero.

Kelly (1973) quotes an expression due to Knowles & Kelly for τ

$$\tau = \frac{\Delta G}{4\pi^2} \left(\frac{3\Delta G}{Gb}\right)^{\frac{1}{2}} \left[0.8 - 0.143 \ln (r/b)\right]^{\frac{3}{2}} r^{\frac{1}{2}} f^{\frac{1}{2}}, \quad \text{for } r > 2b. \Big]$$

$$(2.25)$$

Here ΔG is the difference in shear modulus between matrix and precipitate, and G is the matrix shear modulus.

This expression is yet another in which τ varies with $r^{\frac{1}{2}} f^{\frac{1}{2}}$, so that considerable difficulties exist when attempting to identify a strengthening mechanism in a particular alloy system in which particle shearing occurs. A linear plot of strength versus $r^{\frac{1}{2}} f^{\frac{1}{2}}$ is not a sufficient test of a particular model: the slope of the plot must also be considered. The most effective way of attempting to identify the critical interaction is, as in fig. 2.7, by plotting predicted strength values against experimental strength values, although as we have seen, this may not always lead to the desired discrimination between alternative mechanisms.

2.1.2   *Impenetrable particles – Orowan looping*
We have seen that, for a number of mechanisms, when particles are sheared by dislocations τ increases as r increases because of the increase of F with r. However, (2.1) indicates that as F increases, the breaking angle ϕ decreases, and eventually becomes *zero*. For larger values of F the dislocation bypasses the obstacle rather than passing through it (fig. 2.10) and the bypassing

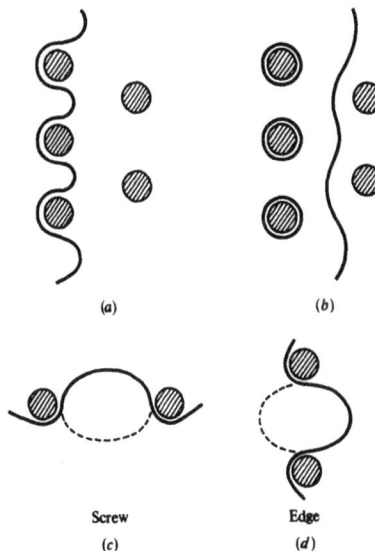

(a)   (b)

Screw   Edge
(c)   (d)

Fig. 2.10. The Orowan process of dislocations (a) bowing between particles, then (b) bypassing the particles by leaving a dislocation loop surrounding each one. (c) and (d) illustrate the bowing between particles of dislocations of screw and edge character respectively.

stress is obtained by substituting $F = 2T$ (2.1) in (2.4), hence

$$\tau = 2T/b\lambda, \tag{2.26}$$

which is the basis of the well-known Orowan equation.

The Orowan equation may be expressed in its simplest form by putting $T = \frac{1}{2} Gb^2$ as

$$\tau = Gb/\lambda \tag{2.27}$$

This equation is, however, often adequate for rough order-of-magnitude calculations of yield-stress increment due to the presence of impenetrable particles of known average spacing. For the purposes of identifying the actual yield mechanism as an Orowan process, however, a more refined version of the equation is required. Such a version may be derived by considering the following parameters.

The mean planar interparticle spacing. As discussed in chapter 1, when calculating the interaction between a gliding dislocation and an array of obstacles in its slip plane, the 'mean planar' separation (λ_S) of the particle should be considered. For a *random* array of obstacles the effective spacing is taken (1.30) as $1.23r (2\pi/3f)^{\frac{1}{2}}$, so the critical flow stress is reduced by a factor of $1/1.23$ (i.e. 0.81) due to this statistical factor. Again, if the size of the particles is not negligible in comparison with their separation, then one must consider the surface-to-surface mean planar separation of the particles, which reduces λ_S by $2r_S$, as shown in (1.32).

Dislocation dipole effect. Ashby has considered the effect of the interaction between the two arms of the bulging dislocation on opposite sides of the particle (fig. 2.10). These dislocation arms have opposite signs and consequently there is an attractive force between them which will reduce the bypass stress. The two arms of the bulging dislocation form a *dislocation dipole* with a width equal to the particle diameter $(2r_S)$, and Ashby calculated that the bypassing stress would be reduced by a factor obtained by writing the logarithmic term in the dislocation line tension equation as $\ln(2r_S/r_0)$, where r_0 is the dislocation inner cut-off ('core') radius, sometimes taken as equal in magnitude to the Burgers vector of the dislocation, b.

The line tension of the dislocation. The line tension of an edge dislocation is greater, by a factor of $(1 - \nu)^{-1}$, than that of a screw dislocation. When a screw dislocation moves as shown in fig. 2.10, the extra length of dislocation that forms as it bows between two particles is mainly of *edge* character, whereas an approaching edge dislocation bows to form mainly *screw* dislocation. At first, therefore, it may appear that the bypass (Orowan) stress for an edge dislocation will be lower than that for a screw dislocation. This is not so, however, for the following reason.

As a dislocation loop expands on the slip plane under the applied stress, the *curvature* of a screw dislocation (fig. 2.10c) will be less than that of an edge dislocation (fig. 2.10d). The mean spacing of obstacles along the screw dislocation will therefore be larger than along the edge dislocation, and this effect compensates for the difference in line tension. In other words, in (2.26) the ratio T/λ is constant for edge and screw dislocations, and the appropriate value of T to be substituted in the equation is the geometric mean value of T for edges and screws.

These three parameters can thus be incorporated in (2.26) to provide a more exact theoretical estimate of the Orowan stress

$$\tau = \frac{0.81 \, Gb}{2\pi(1 - \nu)^{\frac{1}{2}}} \frac{\ln(2r_S/r_0)}{(\lambda_S - 2r_S)} \, . \tag{2.28}$$

Experimental verification of the Orowan relationship

A common tendency in the literature is to plot yield stress against a parameter such as $(\lambda_S - 2r_S)^{-1}$; if a straight line is obtained, it is then concluded that the Orowan equation is obeyed. This conclusion is not necessarily valid unless the experimental values of the *slope* of this straight line are in agreement with those theoretically predicted. Fig. 2.11 illustrates yield-stress data obtained at 77 K with copper single crystals containing dispersed particles of Al_2O_3 or of BeO. The data are plotted in accordance with (2.28), and the straight line obtained has a slope of 90 μm, which may be compared with the theoretically predicted slope of 93 μm. This provides good evidence for these crystals deforming by the Orowan process. Furthermore, the intercept on the τ/G axis of 0.6×10^{-4} agrees well with the value of τ/G of 0.59×10^{-4} for pure copper at the temperature of this test.

It is equally possible to demonstrate the validity of the Orowan equation in describing the yield behaviour of polycrystalline alloys. The 0.2% proof stress is measured $(p_{0.2})$, and the increase in critical resolved shear stress due to the presence of the particles may be obtained by dividing the polycrystalline values of $(p_{0.2} - p_0)$ by the appropriate Taylor factor relating polycrystalline yield stress and critical shear stress in single crystals (p_0 is the 0.2% proof stress in the precipitate-free alloy).

This approach has been made by Chaturvedi, Lloyd & Chung (1976) in the γ'-strengthened Co–Ni–Cr system referred to earlier (see fig. 2.7).

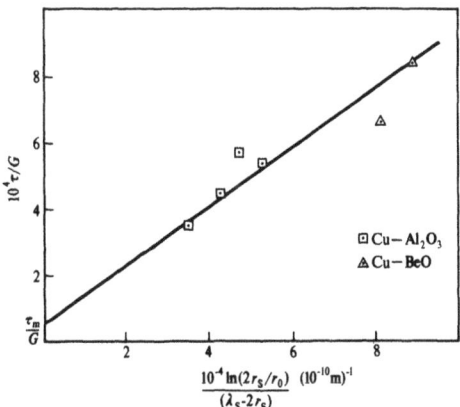

Fig. 2.11. Yield-stress data at 77 K for copper single crystals containing dispersions of Al_2O_3 or of BeO. (After Gould, 1971.)

When these alloys were aged so that the particle size of the Ni_3Ti-phase was greater than 5.6 nm in diameter, Orowan looping rather than particle shearing was observed to take place at yield, when deformed samples were examined by transmission electron microscopy. Fig. 2.12 shows the plot of the observed flow-stress increment due to the presence of the particles compared with that calculated according to (2.28). As can be seen, it is a straight line with a slope of 1.004 ± 0.018, and so it may be concluded that the yield stress can well be accounted for on the basis of the Orowan model. These results are in agreement with those of other workers who have studied the yield stress of stainless steel which has been precipitation hardened by a γ'-phase.

Spheroidized carbides in tempered (ferritic) steels are another family of materials which might be expected to have their yield stress related with their structure by means of an Orowan equation. The problem here is that the dispersion of carbide particles is unlikely to be the only obstacle to dislocation movement. In the quenched condition such steels are in the form of martensite of fine grain size, with a high density of lattice defects such as dislocations. On tempering, a structure consisting of carbides dispersed in a dislocation substructure will be produced, and it is by no means certain what the strength-controlling parameter will be.

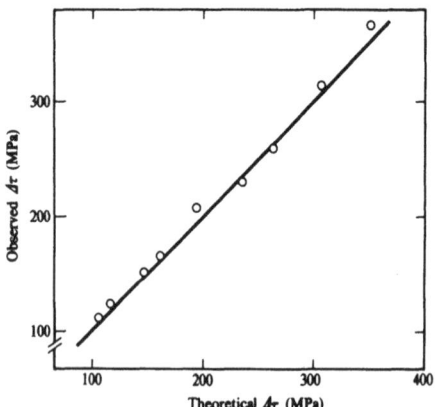

Fig. 2.12. Plot of observed versus theoretical (calculated using (2.28)) values of $\Delta\tau$ during the Orowan process in a Co–Ni–Cr superalloy. (After Chaturvedi, Lloyd & Chung, 1976.)

2.1.3   Dislocation movement through arrays of obstacles of two different strengths

This problem has been considered by Foreman & Makin (1967), who devised a computer programme to determine the stress required to move a dislocation through an array of obstacles arranged randomly on a slip plane. They assumed that the dislocation would break away from an obstacle when the breaking angle ϕ is achieved (2.1), and an array of 10 000 obstacles was used in their computation. The critical shear stress of a crystal containing two types of obstacle of differing strengths was computed over the full range of relative concentrations of both types. Obstacles of each type were distributed randomly, and the results are shown in fig. 2.13 for three different combinations of obstacle strengths. Here, the value of λ is the square lattice spacing, counting all the obstacles irrespective of type.

For strong obstacles ($\phi = 10°$) of a single type, it is seen that the critical shear stress is $0.8Gb/\lambda$, in close agreement with (2.28). For two fairly weak obstacles (lower curve in fig. 2.13) the stress is given very accurately by the root mean square of the stresses for the two arrays taken separately, i.e.

$$\tau^2 = \tau_1^2 + \tau_2^2, \tag{2.29}$$

where τ_1 and τ_2 are the critical shear stresses for the individual arrays of the two types of obstacle.

For two strong obstacles (upper curve of fig. 2.13), the stress lies

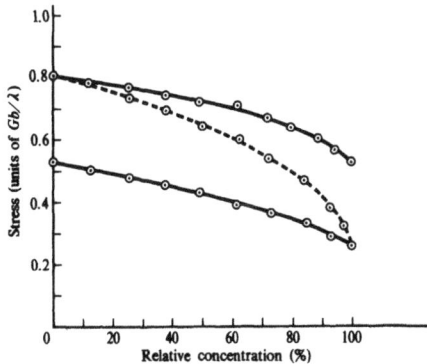

Fig. 2.13. Variation of critical shear stress with relative concentration for random arrays containing 10 000 obstacles of two different strengths, shown for three different combinations of strengths. Breaking angles are 10° and 90° (upper curve), 10° and 130° (middle curve) and 90° and 130° (lower curve). (After Foreman & Makin, 1967.)

between the value given by (2.29) and the arithmetic sum of the separate stresses. With regard to arrays of obstacles of widely differing strengths (middle curve of fig. 2.13), it is interesting to note the large effect of a small concentration of strong obstacles on the stress required to shear an array of predominantly weak obstacles (extreme right of middle curve, fig. 2.13): less than a 10% concentration of strong obstacles increases the stress by 50%. In contrast, a small concentration of weak obstacles has almost no effect on an array of strong obstacles. The arithmetic sum ($\tau = \tau_1 + \tau_2$) is more commonly used in the literature, and is indeed better in the case of few strong obstacles and many weak ones.

Staniek & Hornbogen (1973) classified the basic hardening mechanisms which produce an increase of yield stress by using the geometrical dimensions of the particular obstacle to dislocation motion:

0-dimensional – solid solution hardening (S);

1-dimensional – dislocation hardening (D);

2-dimensional – grain-boundary hardening (B);

3-dimensional – particle hardening (P).

They point out that, if combinations of two are considered (including combinations of the same type, e.g. two different types of particle hardening, PP*), there are ten possibilities. These are given in the table below.

	P	B	D	S
S	SP	SB	SD	SS*
D	DP	DB	DD*	
B	BP	BB*		
P	PP*			

All the mechanisms except those involving grain boundaries can be treated on the basis of the interaction of individual dislocation lines with a mixture of obstacles as discussed above, e.g.

$$\Delta p_{\mathrm{D}} + \Delta p_{\mathrm{P}} > \Delta p > (\Delta p_{\mathrm{D}}^2 + \Delta p_{\mathrm{P}}^2)^{\frac{1}{2}}, \qquad (2.30)$$

for two strong obstacles.

The increase in yield stress due to the two mechanisms ρ_{P} and ρ_{B} can be written as

$$\Delta p = \Delta p_{\mathrm{P}} + \Delta p_{\mathrm{B}} = \Delta p_\rho + k_{\mathrm{Y}} d^{-\frac{1}{2}}, \qquad (2.31)$$

which is a form of the Hall–Petch relation (k_{Y} being the Petch slope). This equation tacitly assumes that Δp_{P} is linearly additive to grain-boundary strengthening, which may not always be true. No adequate theoretical treatment considering large numbers of dislocations is presently available for a more refined treatment of the additivity problem. There appears to be some evidence that either Δp_{P} or $k_{\mathrm{Y}} D^{-\frac{1}{2}}$ can dominate (2.31) in different alloy systems, apparently depending upon the fineness of the particle dispersion.

The yield stress of a single-phase polycrystalline alloy obeys a Hall–Petch equation, since at yield only grain boundaries act as dislocation sources. The grain-size dependence of yield, up to a few % of plastic deformation, can be explained by the fact that for a given amount of plastic deformation, ϵ, the dislocation density generated by the grain boundaries, $\Delta\rho_B$, decreases with increase in grain size. If the average distance of travel of each dislocation, x, is proportional (or equal) to the grain size, d, we can write

$$\epsilon = b\rho_B x = C b \rho_B d,$$

where C is a proportionality constant. Thus

$$\rho_B = \epsilon/C b d.$$

If we now assume that the relationship between dislocation density and flow stress is characteristic of the normal 'forest' hardening model (2.49)

$$\Delta p = \alpha G b\rho^{\frac{1}{2}}$$

where α is a constant, we obtain the yield-stress increment, Δp_B due to the grain boundaries

$$\Delta p_B = \alpha G (\epsilon b/C)^{\frac{1}{2}} d^{-\frac{1}{2}}, \tag{2.32}$$

which corresponds to the Hall–Petch relation.

Equation (2.32) will apply to alloys in which only grain boundaries act as dislocation sources at yield, and where the motion of dislocations is not impeded between the grain boundaries. When second-phase particles are present, a precipitate may, in some circumstances, act as a source of matrix dislocations at the same stress level as that at which dislocations are produced by grain boundaries. Finely-dispersed coherent particles are unlikely to act in this way, but coarser, incoherent particles may generate such dislocations whose density ρ_P will depend upon the volume fraction of such particles present. Furthermore, if noncoherent particles are able to produce dislocations, the path of travel of the dislocations in the interior of the grain, x, will be increasingly impeded by dislocations which have originated in the same way from other neighbouring particles.

Hornbogen & Staniek (1974) examined the superposition of precipitation and grain-boundary hardening in an α-Fe–Cu alloy heat-treated to produce a range of grain sizes and with different dispersions of fcc Cu particles. Their data are shown in fig. 2.14, where it can be seen that the single-phase solid solution obeys a Hall–Petch equation. In the early stages of ageing, when small coherent particles are present (ageing up to 10 hours

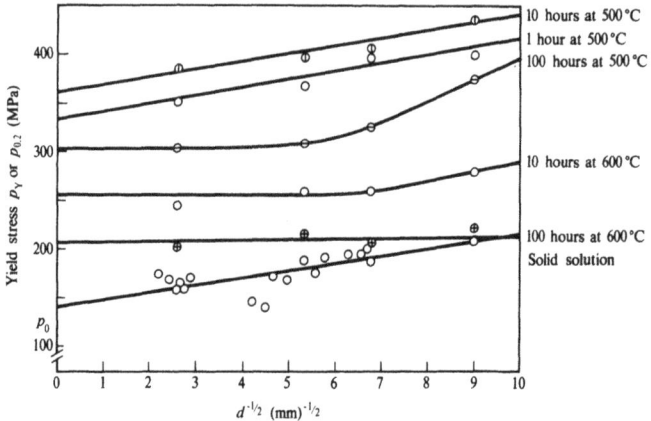

Fig. 2.14. Grain-size dependence of the yield stress of an Fe–1wt% Cu alloy in different precipitation conditions. (After Hornbogen & Staniek, 1974.)

at 500 °C), the data obey (2.31); grain-boundary and particle strengthening are seen to be additive. However, on ageing for longer times or at higher temperatures, the yield stress becomes independent of grain size (although some dependence is apparent when the grain size is fine, after intermediate ageing treatments). Fig. 2.14 thus illustrates that the Hall–Petch relationship is valid in the range of grain sizes in which the average free path of travel, x, is not smaller than the smallest grain size. As x becomes smaller, however, an independence of yield stress from the grain size is observed for all grain sizes larger than x.

An · example of the yield stress being controlled essentially by the particle dispersion is provided by the recent work of Törrönen (1976), who studied a Cr–Mo–V pressure-vessel steel containing 0.18wt% C. The as-quenched microstructure was bainitic, and tempering was carried out at temperatures in the range 600 to 760 °C for various periods up to 90 hours, producing a finely-dispersed precipitate of vanadium-rich MC-type carbides, and a somewhat coarser distribution of M_7C_3 and $M_{23}C_6$ carbides. Törrönen demonstrated that the spacing of the MC-carbides alone determined the change in yield stress with heat-treatment, the coarser particles and the substructure having no influence.

The critical resolved shear stress was calculated from the 0.2% tensile proof stress by dividing by 2.75, and the yield data are plotted in accordance with the Orowan equation (2.28) in fig. 2.15a. Fig. 2.15b illustrates the good agreement between the observed and calculated yield-stress increment.

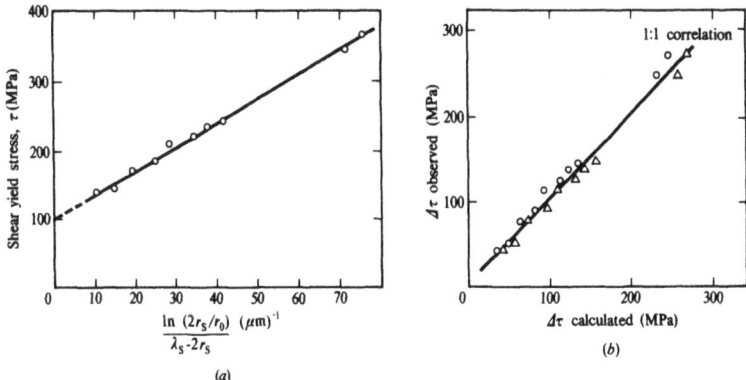

(a)

(b)

Fig. 2.15. Yield-stress data for a Cr–Mo–V pressure-vessel steel tempered to contain MC-type carbides showing correlation with the Orowan process. (After Törrönen, 1976.)

2.1.4   *The yield strength of spheroidized steels*

The role of cementite particles in the strengthening of spheroidized plain carbon steels has been widely studied. The spheroidization of cementite is conveniently achieved by austentizing, quenching and tempering, and Anand & Gurland (1976a) have recently investigated the combined effects of cementite particles, grain boundaries and subgrain boundaries on the yielding behaviour of such steels. For this purpose, spheroidized steels were produced with a range of grain sizes, cementite particle sizes and concentrations, with and without subgrain networks. The yield strengths of such steels were then compared.

The dislocation subgrain network present in such steels is inherited from the shear transformation on quenching. In annealed steels, these polygonized sub-boundaries interact with the dispersion of carbide particles (average radius r), and an equilibrium subgrain size (L) develops whose magnitude will depend upon the particle size r. If F_P is the pinning force per particle exerted on the subgrain boundaries, then L is determined by equating the driving force for grain growth to the pinning force. In fig. 2.16, a subgrain boundary migrating upwards and intersecting a spherical particle is represented. The drag exerted by the particle, resolved vertically, will be given by $\pi r \sigma_{sb} \sin 2\theta$, where σ_{sb} is the specific sub-boundary surface energy (which is effectively equivalent to a surface tension). The maximum value of the pinning force, F_p, will thus be exerted when $\theta = 45°$, where its value will be $\pi r \sigma_{sb}$. The force per unit area of boundary will therefore be $F_p N_S$. But in order to express N_S in terms of N_V, we cannot use (2.12), which assumed a random distribution of particles relative to the boundary. In the present case, all the particles

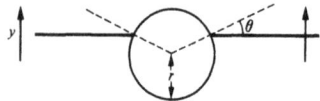

Fig. 2.16. Interaction of a migrating boundary with an inclusion.

lie on the cell boundaries, so if S_V is the boundary area/unit volume

$$N_S = N_V/S_V = N_V L/2.$$

Since $N_V = 3 f/4\pi r^3$, from (1.19), we can write

$$N_S = 3fL/8\pi r^3. \tag{2.33}$$

The driving force for subgrain growth will be given by σ_{sb}/R, where R is the radius of curvature of the subgrain boundaries. If we assume $R = L$, then, at equilibrium

$$N_S F_p = \sigma_{sb}/L. \tag{2.34}$$

Substituting for F_p and N_S we obtain

$$L = C r f^{-\frac{1}{2}},$$

where C is a numerical constant. Allowing for the fact that the condition ($S_V = 0, r = 0$) is not realizable physically, Anand & Gurland obtain

$$L = C' r f^{-\frac{1}{2}} + L_0, \tag{2.35}$$

where L_0 is the lower limit of applicability of the relation. Fig. 2.17 demonstrates that (2.35) is obeyed by experimental data obtained from quenched and tempered steels.

The yield strength of specimens containing cementite particles dispersed upon the dislocation sub-boundaries has been measured. In the presence of the subgrain network, the Orowan yielding mechanism does not appear to operate. It has been suggested that the role of the cementite particles in quenched and tempered steels is indirect, in that these particles stabilize a subcell size which in turn governs the yield stress. Accordingly, the structural parameter which should be used is the mean free path in the ferrite $(\lambda_{l,p})$, which is the cell size, L, corrected for the presence of the second phase,

$$\lambda_{l,p} = L (1 - f).$$

If this parameter is inserted in the Hall–Petch equation, fig. 2.18 illustrates

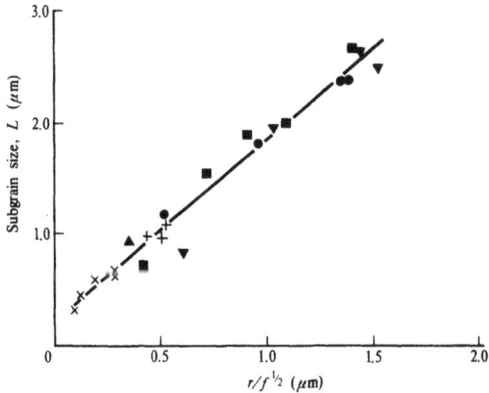

Fig. 2.17. Variation of subgrain size, L, versus the parameter $r/f^{\frac{1}{2}}$, where r is the particle radius and f the volume fraction. (After Anand & Gurland, 1976a.)

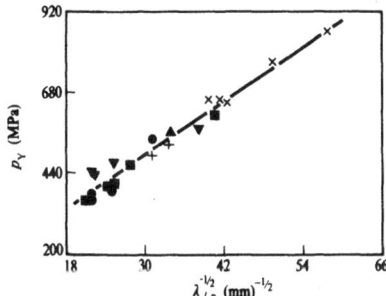

Fig. 2.18. Yield stress p_Y, of spheroidized steels with cementite particles dispersed upon the dislocation sub-boundaries as a function of $\lambda_{l,p}^{-\frac{1}{2}}$. (After Anand & Gurland, 1976a.)

the good fit which is obtained over a wide range of the spacing parameter, $\lambda_{l,p}$.

By subjecting quenched and tempered steels to a thermal-cycling heat-treatment about the A_1 temperature, each cycle was found to result in a decreasing dislocation density, until a relatively 'clean' matrix with a homogeneous dispersion of spheroidal cementite particles lying within the ferrite grains was obtained. The yield strength of this type of structure is found to obey a Hall–Petch relation with respect to the grain size, as seen in fig. 2.19, in contrast to the overaged iron-copper alloy described in fig. 2.14. Thus, in the case of tempered steels with no substructure, the $k_Y d^{-\frac{1}{2}}$ term overshadows the Δp_P term in (2.31).

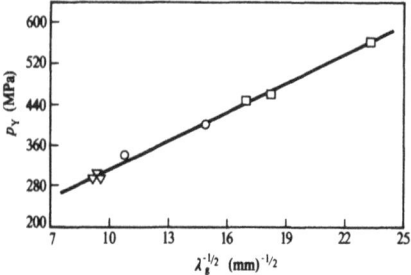

Fig. 2.19. Yield stress, p_Y, of spheroid steels with cementite particles within the ferrite grains as a function of $\lambda_g^{\frac{1}{2}}$. (After Anand & Gurland, 1976a.)

2.2   Work-hardening in two-phase alloys

As in the case of the yield stress, it is convenient to classify the work-hardening behaviour of particle-hardened alloys into two categories: (*i*) that involving shear of the particles by the glide dislocations; and (*ii*) that involving dislocations looping between the particles. In case (*i*), the tensile stress–strain curves of single crystals have essentially the same shape and characteristics as the curves of crystals of pure metals and solid solution alloys. Most alloys containing zones of coherent precipitates belong to this category, and since the theories of work-hardening of single-phase crystals are beyond the scope of this text, we will not consider them further here.

Most alloys containing incoherent particles belong to the second category, and these alloys have single crystal stress–strain curves quite different from those of pure single crystals, or of single-phase solutions. In order to formulate the basis of a theory of work-hardening in the class of alloy involving dislocation bypass of the particles, we must first examine the dislocation structures that develop during the deformation of these materials.

2.2.1   Deformation mechanism in materials containing impenetrable particles

If the particles are small, the dislocation reactions are particularly simple. When the alloy is deformed beyond the yield stress, Orowan loops are formed by the process shown in fig. 2.10 the average number per article, n_{OR}, being given by

$$n_{OR} = 2r\gamma/b, \tag{2.36}$$

where r is the particle radius, b the Burgers vector and γ the plastic shear strain. These are termed the *geometrically necessary* dislocations, because they are required for the compatible deformation of the two-phase system.

As deformation proceeds, n_{OR} increases, and one or more *relaxation mechanisms* occur.

In discussing the type of relaxation mechanism taking place, it is important to distinguish between small and large particles at a given strain. For small particles and small strains, electron micrographs have shown that the first relaxation mechanism to occur is the formation of rows of primary prismatic dislocation loops pressed against the particles (fig. 2.20). The loops are found to be predominantly interstitial in character, and the mean loop diameter is of the same order as the particle radius. When the number of Orowan loops exceeds a critical value, the process illustrated in fig. 2.21 is thought to occur: under the influence of the stress of a glide dislocation the innermost Orowan loop collapses to form two prismatic loops by a double cross-slip mechanism. An interstitial loop and a vacancy loop are formed on opposite sides of the particle. The glide dislocation then bypasses the particle (thus restoring an Orowan loop) then interacts with the second (vacancy) prismatic loop to form a double jog, which is then carried away from the particle. (Only one Orowan loop is shown at the particle in fig. 2.21, for the sake of clarity.) In the case of alloys with a matrix of aluminium or of copper, Orowan loops are not observed, but are likely to anneal out at room temperature, by climb round the particle.

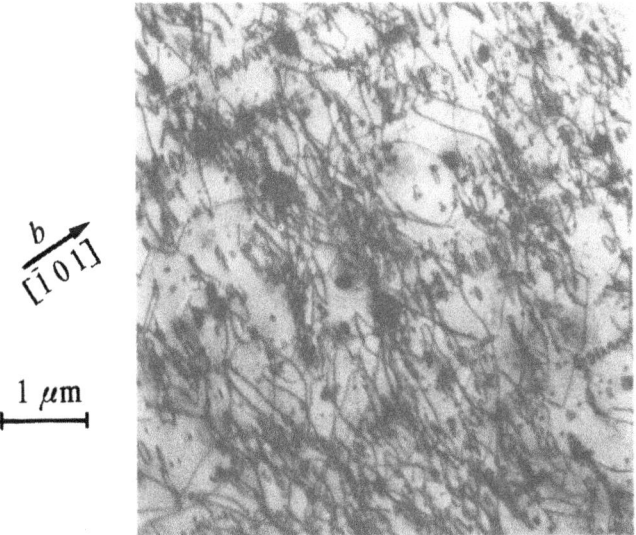

Fig. 2.20. Electron micrograph showing dislocation distribution in the (111) primary slip plane in an aluminium crystal containing silicon particles which has been sheared 6% at 295 K. Alignment of primary prismatic dislocation loops along [$\bar{1}$01], the direction of the Burgers vector, is indicated. (After Stewart & Martin, 1975.)

Fig. 2.21. Schematic representation of the cross-slip process for a dislocation at a precipitate particle. (After Hirsch & Humphreys, 1970.)

As indicated in fig. 2.22, when the matrix is sheared, the particle/matrix interface is subjected to compressive stresses in two quadrants and to tensile stresses in the other two. With increasing particle size, a second form of relaxation, namely *cavitation*, can occur on the tensile side of the particle. Once a cavity is formed, further interstitial prismatic loops are produced by the process indicated in fig. 2.21, the vacancy loop effectively disappearing into the cavity.

The number of Orowan loops needed in a given alloy before cross-slip occurs will depend on the particle size, the presence of any misfit strain associated with the particle, the stacking-fault energy of the matrix, and any back-stress present. The average local shear exerted by an Orowan loop on the particle and material immediately surrounding it is approximately $Gb/2r$, and for n_{OR} loops, using (2.36), this local shear stress is about $G\gamma$. If n_{OR} is taken as 2 for copper alloys containing oxide particles, the critical strain for cross-slip (and prismatic loop formation) to occur is 1%, and the local shear stress due to the loops is about $G/100$. As more Orowan loops transform to prismatic loops, the back-stress from these make further transformation more difficult, and the number of Orowan loops per slip line will therefore increase with strain.

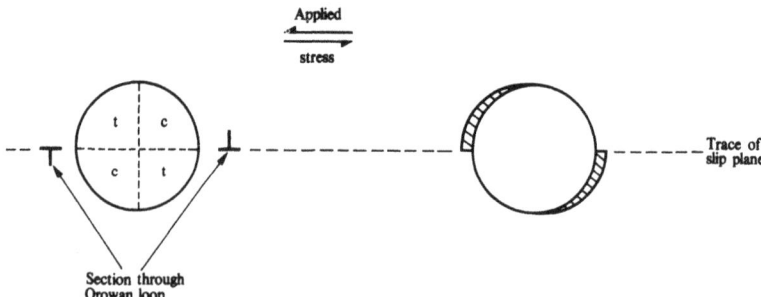

Fig. 2.22. Schematic illustration of tensile (t) and compressive (c) stresses upon a particle/matrix interface when the matrix is sheared. The shaded area represents cavitation.

Other relaxation mechanisms. With increasing strain for a given particle size, the double cross-slip mechanism described above becomes more difficult, because the stack of primary prismatic loops (fig. 2.20) may be unable to glide away from the particle, either by becoming blocked by another particle, or by interaction with secondary dislocations. In solution-hardened alloys, the solute friction stress may also hinder the glide of the prismatic loops. A back-stress from these loops builds up at the particle, and *secondary* dislocations (i.e. of another Burgers vector) are generated to relax the stress at the particle. Thus, at a critical strain it may become easier to relieve the shear stresses by punching out dislocations on another system.

The basic concept of prismatic punching (fig. 2.23) is that a stress concentration at an interface can be removed by creating a dislocation loop at the interface. In doing so, a second prismatic loop is also formed, the stress tending to drive it from the interface. In their study of Cu–20wt% Zn alloys containing SiO_2 particles, Humphreys & Stewart (1972) found that secondary dislocations are punched out at critical strains which decrease with increasing particle size. Fig. 2.24a illustrates the transition from the α-structure (Orowan and prismatic loops of primary Burgers vector $[\bar{1}01]$) to the β- and γ-structures. These latter structures are illustrated in fig. 2.24b. The β-structure consists of the α-structure plus loops of Burgers vector $[110]$ punched out as shown. The γ-structure consists of the β-structure plus prismatic loops punched out with Burgers vectors $[01\bar{1}]$ and $[1\bar{1}0]$ in the primary plane. Humphreys & Stewart also found that cavities were produced with the β-structures, and thus that cavitation occurs at smaller strains for larger particles.

With increasing strain for a given particle size, a further transition occurs to δ- and ε-structures. In the latter, loops with $[101]$ Burgers vector are generated, and in both, some of the prismatic loops become sheared along their glide cylinder, hence causing local lattice rotations. Further increase in strain leads to the formation of very complex structures

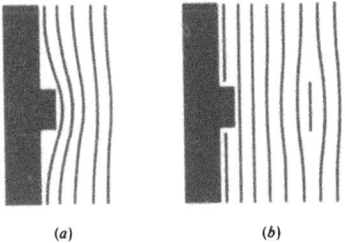

(a)   (b)

Fig. 2.23. 'Punching' of prismatic interstitial dislocation loop into matrix by a particle.

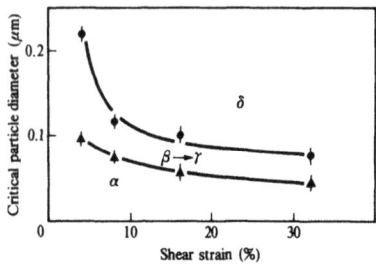

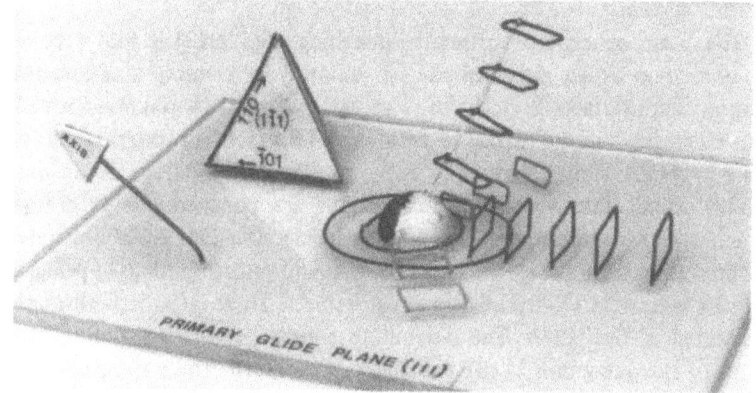

Fig. 2.24. (*a*) Critical particle size for the α to β and γ to δ transitions in structure. (*b*) A wire model for the structure for the top half of α, β and γ structures. A void is present at the dark surface of the particle. (Courtesy of F.J. Humphreys.)

produced by the interaction of the secondary dislocations with one another and with primaries. A similar pattern of behaviour has been observed by Stewart & Martin (1975) in an Al–Cu matrix containing particles of Si. In this system, however, cavitation at the particle/matrix interface was not found, so relaxation occurred by the generation of loops with [011] Burgers vector, which relax the stress in those quadrants in which the interface is in a state of tension (fig. 2.22).

The loops punched out on secondary systems will in their turn increase the back-stress on these systems until it becomes favourable for the primary cross-slip relaxation process to occur again. A dynamic equilibrium is thus set up between the various relaxation processes, and it is clear that plastic zones of considerable complexity are produced.

The effect of high strains

When considering the deformation structure produced at high strains adjacent to large particles, it is obvious that the complexity of the problem precludes a consideration of the movement of individual dislocations. A more macroscopic continuum approach has been made by Humphreys (1977) which can account for the structures observed in the electron microscope in such deformed material. Fig. 2.25a illustrates a single crystal matrix deforming on a single slip plane AB intersecting a particle. If the matrix deforms by a vector nb, fig. 2.25b shows the particle deforming with the matrix. If the particle is non-deformable, then local plastic deformation must restore the particle to its original shape, leaving the matrix sheared. One way of effecting this shape change is to produce a local reverse shear of nb across the plane CD by locally rotating a portion of the matrix in the sense shown in fig. 2.25c. In three dimensions, rotation of a suitable portion of the lattice about *any* axis normal to b can produce the required shear across CD. As the rotations are accomplished by dislocations, however, the rotation axes depend upon the available slip systems of the crystal.

Arrays of Orowan loops (fig. 2.25d) are unstable owing to their high energy and the high local stresses produced in the matrix, as already discussed. One way of effectively neutralizing the Orowan loops is to

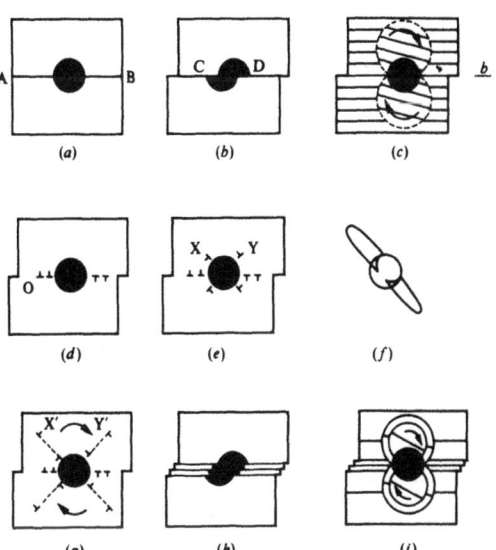

Fig. 2.25. The origin of a deformation zone at a particle. (After Humphreys, 1977.)

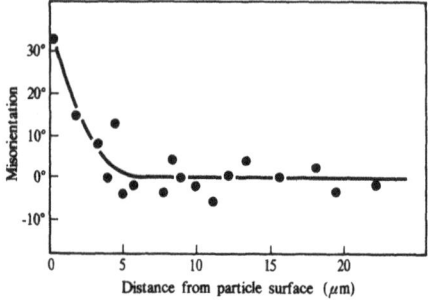

Fig. 2.26. The orientation of the matrix in the vicinity of a 4 μm diameter particle in cold-rolled aluminium. (After Humphreys, 1977.)

produce other loops of opposite sign. Loops such as X and Y in fig. 2.25e can thus effectively cancel out the Orowan loops. Creating such secondary loops at the interface by glide processes requires the creation of additional outer loops as shown schematically in fig. 2.25f, resulting in dislocations X' and Y' in the matrix (fig. 2.25g). The sign and configuration of these dislocations is such that a subgrain with the same type of rotation as in the continuum model (fig. 2.25c) is produced.

In practice, several slip planes intersect a particle (fig. 2.25h), but this can be taken into account on the continuum model by producing a series of inner subgrains with progressive misorientations (fig. 2.25i). Humphreys points out that the magnitude of the rotation θ is related to the shear strain γ, the particle diameter d and the radius of the rotated zone (R) by

$$\theta = \gamma d / 2R. \tag{2.37}$$

Experimentally it is found that R is approximately equal to $d/2$ and thus $\theta \approx \gamma$. However, if several regions rotate about different axes, the rotations about any one axis will be less. For the case of fcc metals where three axes have been identified, it might be expected that $\theta \approx \gamma/3$. Humphreys examined the structure of Al–Cu solid solution crystals containing Si particles of a few microns diameter which were subjected to large plastic strains by cold rolling deformation with more than 50% reduction in thickness. A high dislocation density in the form of subgrains was observed, and by means of selected area electron diffraction, local misorientations adjacent to the particles were detected. Fig. 2.26 illustrates the change in orientation on the matrix adjacent to a particle, where it is apparent that a structure similar to that illustrated in fig. 2.25i has been obtained.

Summary

We can summarize the deformation structures in these plastically inhomogeneous materials in the following way:

For small particles and small strains Orowan loops are left around the particles. At larger strains (probably not exceeding 1 to 2.5% for copper alloys at low temperatures) and small particles, the innermost of these loops then cross-slip to form primary prismatic loops. This leads to the generation of rows of primary prismatic loops aligned with the particles in the direction of the Burgers vector (fig. 2.20). At even larger strains and for larger particle sizes, the back-stress from the prismatic loops raises the stress at the interface to that necessary to generate loops of secondary Burgers vector, and in some systems a void may form at the particle/matrix interface where it is in tension. Loops are punched out on secondary systems, a dynamic equilibrium is set up between the various relaxation processes, and plastic zones of considerable complexity are produced. At the highest strains, large lattice misorientations develop in the vicinity of the particles, i.e. at a distance less than the particle diameter.

2.2.2   Theories of work-hardening

In view of the existence of the various types of dislocation structures which develop as the plastic strain increases, as summarized above, it is necessary to calculate the work-hardening due to a given type of structure, and to attempt to combine the contributions in order to explain the work-hardening from Orowan loops, from primary prismatic loops, and from secondary dislocations.

Work-hardening due to Orowan loops

For small deformations, the contribution from Orowan loops is particularly important. Such loops are thought to determine the Bauschinger effect, to control the strong temperature dependence of the work-hardening (see chapter 4), and to contribute to the recovery observed in the same temperature range. During straining, the gliding dislocations will bow between the particles surrounded by Orowan loops (while themselves forming half-Orowan loops, see fig. 2.10), and the flow stress will be given by

$$\tau_a = \tau_{OR} + \tau_b + \tau_{LR} + \tau_{im} \qquad (2.38)$$

where τ_a is the applied shear stress, and τ_{OR} is the Orowan stress. τ_b is the 'bowing stress' which increases with strain due to the reduction in effective gap between neighbouring particles caused by the repulsion of the bowing glide dislocation by the uniformly distributed Orowan loops left around the particles at smaller strains. τ_{im} is an 'image stress' which acts on the glide dislocation and it arises from the *distant* Orowan loops, and is believed to be almost uniform throughout the crystal. τ_{LR} is another long-range stress from the accumulated Orowan loops, and is relatively uniform over a large area between the particles.

Hazzledine & Hirsch (1974) have recently calculated the shape of the stress–strain curve on a model of coplanar Orowan loops by considering the values of the individual stresses in (2.38). The minimum stress required to propagate dislocations through the alloy is the Orowan stress

$$\tau_{OR} = Gb/\lambda, \tag{2.39}$$

where λ is the average gap between the particles, i.e. $\lambda = D_S - \frac{1}{2}\pi r$, where D_S is the average planar spacing of the particle centres and r the particle radius. $D_S = r(2\pi/3f)^{\frac{1}{2}}$, where f is the volume fraction of the dispersion. As the Orowan loops build up, the gap between the particles decreases and λ becomes $D_S - 2r\phi$, where ϕ is a measure of the radius of the outermost dislocation of the planar array surrounding each particle, being defined as

$$\phi = \bar{r}_n/r,$$

where $\bar{r}_n$ is the average radius of the nth loop at a spherical particle of radius r. Since $2r\phi < D_S$, the stress required to bow between the obstacles may be written as the sum of the Orowan stress and the bowing stress, τ_b, where

$$\tau_b = 2r\tau_{OR}(\phi - \tfrac{1}{4}\pi)/(D_S - \tfrac{1}{2}\pi r). \tag{2.40}$$

Hazzledine & Hirsch write this expression in the form

$$\tau_b = 2\tau_{OR}\, n_{OR}\, F/(D_S - \tfrac{1}{2}\pi r), \tag{2.41}$$

where n_{OR} is the number of Orowan loops, and F is independent of n_{OR}.

The accumulation of Orowan loops leads to the appearance of the two long-range stresses, τ_{im} and τ_{LR}. Brown & Stobbs (1971) have calculated the value of τ_{im} as follows. For simplicity, let the elastic constants of matrix and particles be identical, then when the matrix has undergone a (symmetrical) plastic shear strain γ_p, elastic strains of the order γ_p are induced in the particles. In the absence of any such elastic constraint, plastic flow produces no internal stress, it simply changes the shape of the holes in which the particles sit. However, in the present case, stresses and strains will develop in and around the particles, resulting from the elastic reaction of the material to the plastic flow. A theory due to Eshelby (1957) has been used to calculate the resulting stresses and strains in the crystal. The result is known as the 'image stress' field, since it can be thought of as arising from an external distribution of 'image' inclusions, each cancelling appropriate stresses on the external surfaces of the specimen (in the same way that one can treat the problem of the stresses acting upon a dislocation near the free surface of the crystal by using the concept of an 'image dislocation' whose position is such that it appears to be a reflection of the given dislocation in the specimen surface).

Brown & Stobbs show by the method of continuum mechanics that τ_{im} takes the value

$$\tau_{im} = A \, G f \, \gamma_p, \tag{2.42}$$

where A is an accommodation factor which relates the strain in the inclusion to γ_p and is of order unity. In order to substitute its value into (2.38), Hazzledine & Hirsch express γ_p in terms of the number of Orowan loops, n_{OR}. If the slip line spacing is h, then a fraction $2r/h$ of the total number of particles per volume $(3f/4\pi r^3)$ are surrounded by n_{OR} Orowan loops of area $2\pi n_{OR} \bar{g} r^2 /3$, where $\bar{g}$ is a measure of the mean area enclosed by a loop in a pile-up. The strain is therefore $bfn_{OR} \, \bar{g}/h$ and the image stress is given by

$$\tau_{im} = G \, b f \bar{g} \, n_{OR}/h. \tag{2.43}$$

The long-range stress from the accumulated Orowan loops is estimated by integrating the stress from a set of infinitesimal loops with planar spacing D_S, and is expressed by Hazzledine & Hirsch as

$$\tau_{LR} = k' Gb f \bar{g} \, n_{OR} D_S, \tag{2.44}$$

where

$$k' = 0.38 \, [(2 - \nu)/4(1 - \nu)].$$

The flow stress is thus written by substitution in (2.38):

$$\tau_a = \tau_{OR} + \frac{2\tau_{OR} \, Fn_{OR}}{D_S - \frac{1}{2}\pi r} + \frac{k' Gbf\bar{g} \, n_{OR}}{D_S} + \frac{AGbf\bar{g} \, n_{OR}}{h} \tag{2.45}$$

Fig. 2.27 shows a computed stress-strain curve of an alloy consisting of a copper matrix containing particles with the same shear modulus as copper ($G = 4630 \text{ kg mm}^{-2}$, $\nu = 0.369$, $f = 5 \times 10^{-3}$. $r = 100b$, $D_S = 2050b$), assuming a slip-line spacing $h = 2r$ and $\gamma = n_{OR} \, b/2r$. The values of F and $\bar{g}$ were evaluated for loops under an effective stress $\tau_a - \tau_{im} - \tau_{LR}$. Also shown in fig. 2.27 is the image stress alone, which may be seen to be responsible for the concave shape of the stress-strain curve. It is clear that the theory predicts essentially linear hardening, whereas the experimental curves have continuously decreasing hardening rates. The example shown in fig. 2.27 is for copper with Al_2O_3 particles.

This discrepancy between theory and experiment led Hazzledine & Hirsch to conclude that large coplanar arrays of Orowan loops are unstable, and they develop a 'hybrid' model in which the first few (say four) loops at any particle are Orowan loops and further Orowan loops are assumed to transform into stacks of prismatic loops. They conclude that the work-

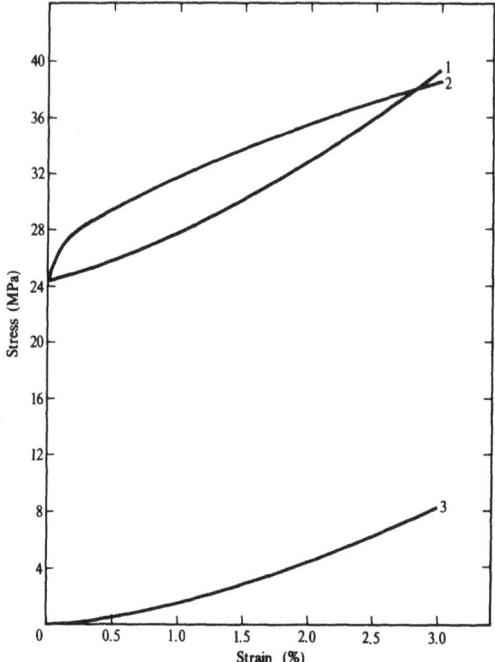

Fig. 2.27. Calculated (1) and experimental (2) stress–strain curves. The calculation is based on a coplanar model of Orowan loops, and curve (3) is the image stress component of (1). (After Hazzledine & Hirsch, 1974.)

hardening is due to Orowan loops at low strains and, to an increasing degree, due to prismatic loops at high strains.

Work-hardening due to prismatic loops

A theory for this mechanism has been developed by Hirsch & Humphreys (1970), in which the hardening of a slip line is due to the rows of prismatic loops pressed against the particles (fig. 2.20). These rows act as linear obstacles to the gliding dislocations, the length of the obstacles increasing with increasing strain.

In the theory discussed in the preceding paragraph, the slip-line spacing has been assumed to be constant, and it has also been assumed that there is one slip per line per particle. It is possible, however, that the slip-line spacing varies with the strain, and this will, of course, change the shape of the stress–strain curve. By minimizing the flow stress for a given strain, the variation of slip-line spacing can be derived. This theory was, in fact, used by Hirsch & Humphreys for the hardening arising from prismatic loops.

The theory predicts that the slip-line spacing h is proportional to $\gamma^{-\frac{1}{2}}$, and that the number of prismatic loops per slip line n_p ($= n - n_{OR}$) is proportional to $\gamma^{\frac{1}{2}}$. The presence of n_{OR} Orowan loops per particle will add the image stress hardening term previously discussed, and the Hirsch & Humphreys expression may be written

$$\tau_a = \tau_{OR} + \frac{q_1 T(n - n_{OR})r}{b D_S^2} + \frac{q_2 Gb}{4\pi K h} + \frac{A G b n_{OR}}{h}, \qquad (2.46)$$

where q_1 is a numerical factor which depends upon the breaking angle ϕ (for $\phi = \frac{1}{2}\pi$, $q_1 = 1.64$), T is the dislocation line tension, K is 1 for screws and $2(1 - \nu)$ for edges, and q_2 is a factor which takes into account the fact that when reaching the critical breaking angle at the row of loops, the dislocation may not have to overcome the *maximum* passing stress. Equation (2.46) may be written

$$\tau_a = \tau_{OR} + \left(\frac{q_1 q_2^{\frac{1}{2}} GTr}{\pi K b D_S^2}\right)^{\frac{1}{2}} (\gamma^{\frac{1}{2}} - \gamma_0^{\frac{1}{2}}), \qquad (2.47)$$

where

$$q_2^{\frac{1}{2}} = q_2 + 4 K A f n_{OR},$$

and γ_0 is the strain above which the prismatic loop hardening is dominant. Equation (2.47) thus predicts a parabolic work-hardening curve.

Work-hardening due to secondary dislocations

As described in §2.2.1, with increasing strain and/or increasing particle size, a complex plastic tangle is formed around the particles. The work-hardening in this case can be considered to be partly due to the *forest hardening*[†] in the plastic zones, since a gliding dislocation will encounter these as well as the particles and dislocation debris around the particles in the primary slip planes.

This problem has been considered by Ashby (1971), who obtained a rough estimate of the short-range interaction between an array of stored dislocations and a gliding dislocation. He described the array by one parameter only, its average density. Ideally, a work-hardening theory should include the detailed structure of the array, but a useful analysis emerges on the simpler basis which may be applied not only to model systems but also to describe the work-hardening behaviour of certain polycrystalline particle-hardened systems. This is perhaps not surprising,

---

[†] Forest hardening arises from the interaction between a gliding dislocation and other dislocations *not* lying in the primary slip plane: the gliding dislocation thus 'sees' them as one sees the trunks of trees in a forest.

since in such materials multiple slip will occur due to the normal complexity effect in polycrystals, so that complex plastic tangles are likely to be associated with the particles from the start of plastic deformation.

Ashby derives a relation between the flow stress and the dislocation density by means of a simple dimensional analysis. The flow stress, τ, must depend on an elastic constant, G, and the magnitude of the Burgers vector in the material, b. The dimensionless combination of these terms may be written

$$\tau/G = \alpha(b^2 \bar{\rho})^m, \tag{2.48}$$

where both sides are pure numbers, and α and m are constants. The applied stress exerts a force τb per unit length of the gliding dislocation; its motion is opposed by the nearest dislocations. The force of interaction between any two dislocations is proportional to b^2, thus τ is proportional to b, which implies that $m = \frac{1}{2}$ in (2.48). In general a frictional stress and long-range internal stresses will also oppose the motion of a glide dislocation, so these can be represented by the term τ_0:

$$\tau = \tau_0 + \alpha G b \, \bar{\rho}^{\frac{1}{2}}. \tag{2.49}$$

A detailed calculation provides a numerical value for the dimensionless constant α of between 0.15 and 0.5.

As mentioned in §2.2.1, if in a two-phase alloy one component deforms plastically more than the other, then the gradients of deformation which are formed require that dislocations be stored. It is the arrangement and density of these geometrically necessary dislocations which determines the extra work-hardening of such alloys. Such dislocations are not required in the uniform deformation of, for example, a pure single crystal. Such a crystal will work-harden by the accumulation of dislocations which results from chance encounters in the crystal leading to mutual trapping; these are sometimes called 'statistically stored' dislocations. The form of the array of geometrically necessary dislocations depends on whether the particles are equiaxed, or plate- or needle-like.

Equiaxed particles. Considering first the shape of equiaxed particles, fig. 2.28a illustrates a cube-shaped cell of matrix containing a single cube-shaped particle. In order to estimate the density of geometrically necessary dislocations we deform the cell uniformly after removing the particle from the cell (fig. 2.28b). In order to replace the particle into the hole from which it came, the hole must be deformed back into its original shape, and many possible sets of displacements can accomplish this. The simplest method is to insert a number of shear loops of dislocations (i.e. Orowan loops) as shown in fig. 2.28c. These have the effect of shearing the

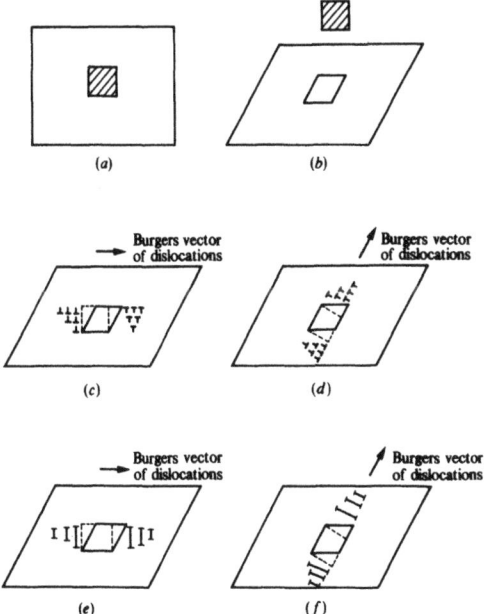

Fig. 2.28. (*a*) An element of crystal containing a hard inclusion of a second phase (shown prior to deformation); (*b*) the inclusion has been removed and the crystal sheared uniformly; (*c*) and (*d*) show the hole restored to its original shape by the emission of glide dislocations on (*c*) the primary system and (*d*) a secondary system; (*e*) and (*f*) show the hole restored to its original shape by prismatic glide.

part of the matrix immediately surrounding the hole back by an amount γ. An array of primary prismatic dislocation loops can have the same effect, as shown in fig. 2.28*d*. The volume ΔV of material which must be removed by prismatic glide is

$$\Delta V = \tfrac{1}{2} V_p \gamma,$$

where V_p is the volume of the particle. This volume has to be injected into the matrix as 'interstitial' prismatic loops, whose Burgers vector is the same as that of the original applied shear strain. On the right-hand side of the particle an equal volume has to be *supplied* from the matrix, which is equivalent to generating 'vacancy'-type prismatic loops.

Many other patterns of prismatic glide are able to overcome the compatibility problem, some not involving loops of primary Burgers vector (fig. 2.28*f*). When a single active slip plane intersects the particle, the array of shear or primary prismatic loops will appear as shown in figs. 2.28*c* and *e*. The loops may not all be of the same size, but geometry requires that their average area be close to $2r^2$, where r is the particle radius. Regardless of the mode of generation of the prismatic loops, the total number of loops per particle (interstitial plus vacancy) n is given by

$$n = V_p \, \gamma/2r^2 b = 4r\gamma/b. \tag{2.50}$$

The total number of loops per unit volume, N, is equal to nN_V. Replacing N_V by $3f/4\pi r^3$ we obtain

$$N = 3\gamma f/\pi b r^2. \tag{2.51}$$

The average diameter of these loops is $\sqrt{2}r$, so a crude measure of the dislocation density is thus $\sqrt{2}\pi r N$, or

$$\rho_G \approx (f/r) \, 4\gamma/b. \tag{2.52}$$

Plate-like particles. Consider a two-phase structure consisting of an array of strong plates spaced a distance D apart in a ductile single crystal matrix (fig. 2.29). The composite is now sheared by slip on a single set of slip planes in a direction normal to the plate surface (fig. 2.29b), so that the central part of each slip plane shears by an amount γ. Since there is zero shear in the matrix immediately adjacent to a plate, the plates are rotated through an angle $\phi = \tan^{-1} \gamma$ relative to the original slip plane, as shown in fig. 2.29b. The matrix lattice thus acquires a curvature of mean magnitude $2\phi/D$. A 'Burgers circuit' construction may be used to calculate the density of geometrically necessary dislocations in the bent matrix crystal, as indicated in fig. 2.29c. Such a circuit has been constructed along BB'NN', which has an area of one-half the area DL of the matrix crystal. Since the closure failure is ϕL, the circuit encloses a number $\phi L/b$ of dislocations each of Burgers vector b. If fig. 2.29 represents a crystal of unit thickness perpendicular to the page, the volume average dislocation density in the crystal enclosed by the circuit is therefore

$$\rho_G = (2/DL) \, (\phi L/b).$$

The angle ϕ can be replaced by γ to a good approximation, giving

$$\rho_G = 2\gamma/Db. \tag{2.53}$$

As pointed out by Ashby, the density of these dislocations can be very large: plates spaced 1 μm apart require a density of about 10^{15} m^{-2} of geometrically necessary dislocations to accommodate the lattice curvature resulting from only 10% shear strain. This is much larger than the density of 10^{12} to 10^{13} m^{-2} of dislocations which are 'statistically' stored in a single-phase copper crystal at a similar strain. These geometrically necessary dislocations can obviously be made to dominate the total dislocation

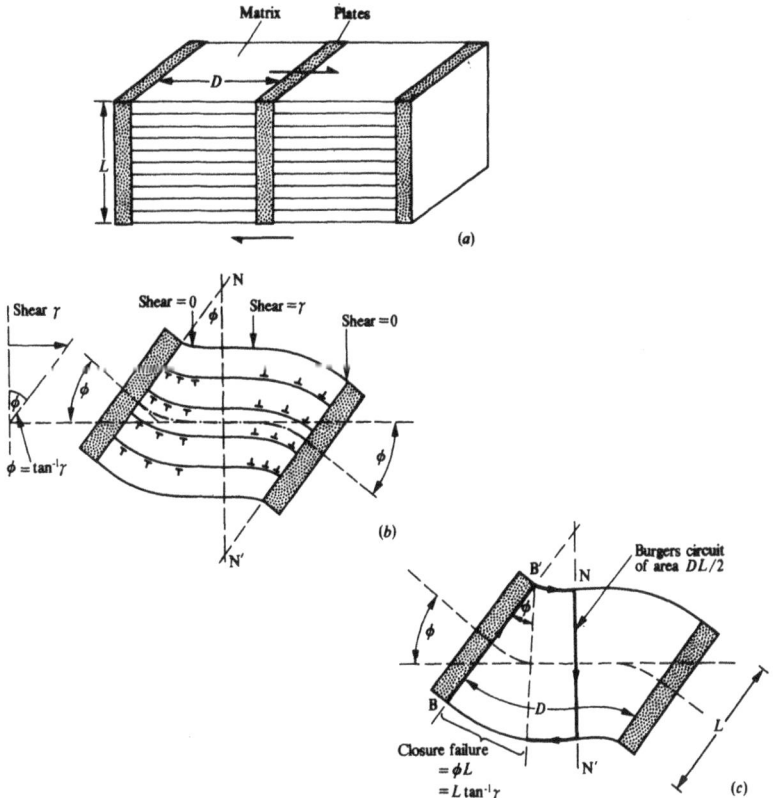

Fig. 2.29. (a) Part of an idealized composite consisting of non-deforming plates strongly bonded to a deformable, single crystal matrix. (b) The ductile matrix is sheared through γ. (c) Closure failure (= ϕL) of a Burgers circuit composed of lattice vectors normal to the slip plane and in the slip direction.

density by making D small enough, see (2.53), either by putting the plates close enough together, or, in the case of equiaxed particles (2.25), by making f/r large enough (i.e. a sufficiently high volume fraction of small particles).

Stress-strain equations. By substituting the appropriate expression for the density of the geometrically necessary dislocations, i.e. (2.52) or (2.53), into the 'forest hardening' equation (2.49), one obtains an equation predicting a parabolic stress-strain curve. Thus for equiaxed particles we obtain

$$\tau = \tau_0 + \alpha' G \left(f b\gamma/2r \right)^{\frac{1}{2}}. \tag{2.54}$$

It is clear that both the size and the volume fraction of the particles influence the flow stress.

For plate-like particles, whose size is comparable with their spacing D, the equation is

$$\tau = \tau_0 + \alpha'G \, (b\gamma/D)^{\frac{1}{2}}, \qquad (2.55)$$

and the work-hardening increases as D, the plate spacing, decreases.

A more complex expression than (2.54) may be derived for the case of equiaxed particles, if one considers the flow stress as arising from the several contributions expressed in (2.38). A gliding dislocation encounters not only the 'forest' of geometrically necessary secondary dislocations emanating as a plastic zone from every particle, but also the particles and dislocation debris remaining around the particles in the primary slip planes. If γ_0 is the unrelaxed strain, one may write

$$\tau = \tau_0 + \alpha G \, (fb/2r)^{\frac{1}{2}} \, (\gamma - \gamma_0)^{\frac{1}{2}} + G\gamma_0 Af, \qquad (2.56)$$

where the last term represents the image stress. With $\gamma \gg \gamma_0$, the second (forest) term is equivalent to Ashby's forest theory.

2.2.3   Analysis of work-hardening data

Single crystal data

In order to test the theoretical predictions outlined above, analysis of stress–strain curves obtained from single crystals containing non-deforming particles has been carried out. For example, Russell & Ashby (1970) have tested two-phase crystals of aluminium containing plate-like particles of $CuAl_2$ and they have also replotted the data of Dew-Hughes & Robertson (1960). Their work-hardening data are plotted according to (2.55) in fig. 2.30, τ_0 being taken equal to the critical shear stress. The line does in fact

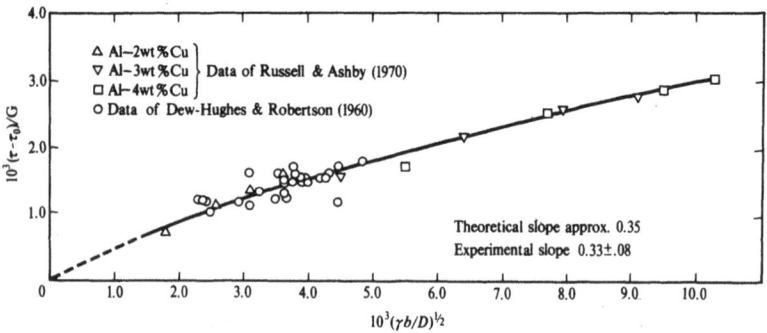

Fig. 2.30. The increment in flow stress due to work-hardening single crystals of aluminium containing plate-like inclusions of θ'-phase and $CuAl_2$, plotted according to (2.55). (After Russell & Ashby, 1970.)

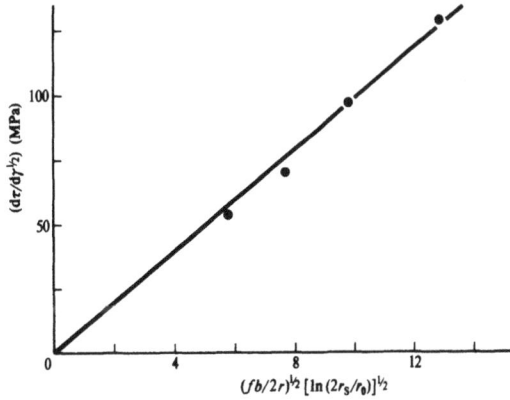

Fig. 2.31. The work-hardening rates of Cu + Al_2O_3 crystals deformed at 77 K plotted to test (2.47) with $T = (Gb/4\pi K') \ln (2r_S/r_0)$ and assuming the particles are spherical so that $1/D_S^2$ is given by $3f/2\pi r$. (After Hirsch & Humphreys, 1970.)

extrapolate to the origin, and has a slope which is in good agreement with theory.

With regard to crystals containing equiaxed particles, work-hardening due to arrays of prismatic loops is predicted by the theory of Hirsch & Humphreys (2.47) to give rise to a parabolic work-hardening curve. Fig. 2.31 illustrates the work-hardening rates observed in copper crystals containing dispersions of Al_2O_3 deformed at 77 K, with the data plotted to test this theory. It is clear that good agreement is obtained with (2.47) if $q_2 \approx 1$.

Moderately good agreement of experiment with theory is, in fact, obtained using Ashby's simple 'forest hardening' approach, (2.54). Fig. 2.32 illustrates the work-hardening at 77 K plotted according to this relationship for copper single crystals containing particles of BeO. The data refer to flow stresses corresponding to a strain of 10%, and are replotted by Ashby from the results of Jones & Kelly. The experimental slope of 0.16 ± 0.05 may be compared with the theoretical slope of approximately 0.25, and the discrepancy presumably arising from the neglect of the other contributions to the flow stress as indicated in (2.56).

Polycrystal data

An appropriate practical system to which the above theories might be applied is that of quenched and tempered steels containing spheroidized carbide particles in a ferritic matrix. In addition to studying the yield behaviour of such steels (see §2.1.4), Anand & Gurland (1976b) have recently examined the strain-hardening behaviour of two high-carbon

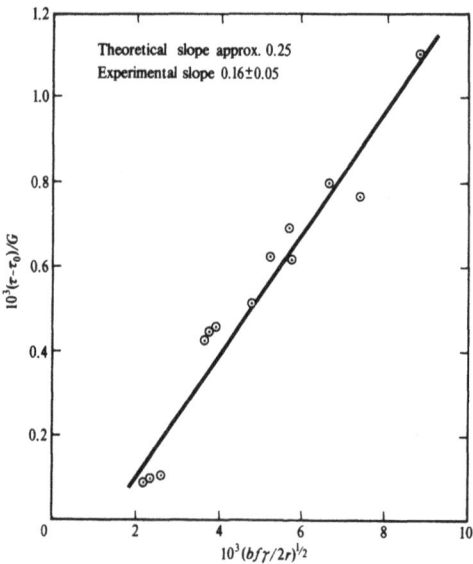

Fig. 2.32. The increment in flow stress after 10% shear strain in single crystals of copper containing particles of BeO deformed at 77 K plotted according to (2.54). (After Ashby, 1971.)

steels heat-treated to produce differing cementite particle sizes. The dislocation substructure was removed by a thermal cycling heat-treatment, resulting in a structure consisting of relatively uniform dispersions of spheroidal cementite particles lying within the ferrite grains, the grain size being greater than the interparticle spacing.

Fig. 2.33 shows the data of Anand & Gurland plotted according to (2.54) showing the tensile flow-stress increment (Δp_p) for (tensile) plastic

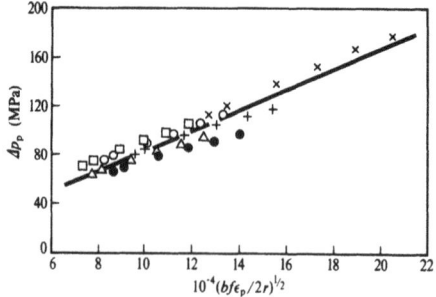

Fig. 2.33. The tensile flow-stress increment, Δp_p, versus the parameter $(bf\epsilon_p/2r)^{\frac{1}{2}}$ for a number of spheroidized steels at plastic strain of up to 3.5%. (After Anand & Gurland, 1976*b*.)

strains (ϵ_p) of up to 3.5%. The hardening behaviour up to 3.5% strain is described by

$$\Delta p_p = 0.11 + 8247 \, (bf\epsilon_p/2r)^{\frac{1}{2}}. \tag{2.57}$$

The very small value of the intercept reflects the contribution to the flow stress of the long-range internal stresses and the statistically stored dislocation density. Taking the shear modulus G as 82.26 GPa, α' in (2.54) is approximately unity, which is in agreement with Ashby's theory.

At plastic strains above 3.5%, however, Δp_p does not continue to increase parabolically with strain, in disagreement with Ashby's theory. As the specimens are strained, dislocation cell structures of dimensions approximately equal to the mean interparticle spacing are formed. Although the mechanisms by which such cells form are not clearly understood in detail, it is clear that the formation of such cells containing groups of dislocations of opposite signs allows for some *recovery*. Thus, interactions between the various dislocations in the cell walls do not permit a continued increase in the dislocation density stored for geometrical reasons. This 'dynamic recovery' will therefore lead to a deviation in the strain-hardening behaviour from that predicted by the Ashby model.

2.3   Plastic flow in two-phase materials with coarse microstructures

Theories of dispersion-strengthening cannot be applied to the commonly encountered microstructures consisting of coarse, hard constituents embedded in a continuous matrix. By 'coarse', one refers to the size of the inclusions being of the order 1 to 100 μm, with volume fractions exceeding 0.1. Aggregates of pearlite in ferrite, as found in normalized plain carbon steels, would be in this category, and recent work in Sweden by Karlsson and his associates has attempted to account theoretically for the properties of this type of microstructure.

In steel with ferrite-pearlite structure, the ferrite is continuous at carbon contents up to 0.4 to 0.5wt%. The pearlitic areas in the microstructure can be regarded as macroscopically homogeneous. Therefore, it should be possible to describe the properties of the aggregate in terms of the plastic properties of the individual constituents, if it were known how strain and stress are distributed between the constituents, and how they interact during deformation. A similar approach can be made to account for the properties of composites such as tungsten carbide dispersed in cobalt, and in Fe-Ni-C alloys consisting of islands of martensite in a matrix of austenite.

The *yield stress* of such aggregates (e.g. of ferrite-pearlite with a continuous ferrite matrix) is found to be about the same as that extrapolated from single-phase matrix data of equal grain size, and is independent of the hardness of the included constituent. However, the *flow stress* and work-hardening rate are considerably higher than that of the single-phase matrix, and rise with increasing inclusion hardness.

During plastic deformation of the aggregate, both constituents deform, the harder one deforming elastically at first but then plastically. The flow stress of the *matrix* can be satisfactorily described in terms of geometrically necessary dislocations, which accommodate the strain difference between the two constituents of the aggregate. In order to estimate the flow stress of the *aggregate*, however, both the stress and strain distribution in the aggregate must be known. Two simple limiting continuum models will provide an idea of the influence of hard inclusions on the stress–strain behaviour of an aggregate. The phases are arranged (fig. 2.34) so that they suffer either the same strain or the same stress (mutual interactions between the phases by constraints, or different lateral strains are neglected).

Yielding starts at the normal yield stress of the soft phase. However, the work-hardening rate is higher than that of this soft single-phase at the same strain. In the equal-strain model (fig. 2.34*a*) this can be ascribed to an increasing stress transfer to the harder phase, which is still elastic. In the equal-stress model (fig. 2.34*b*) the stress in the soft phase increases owing

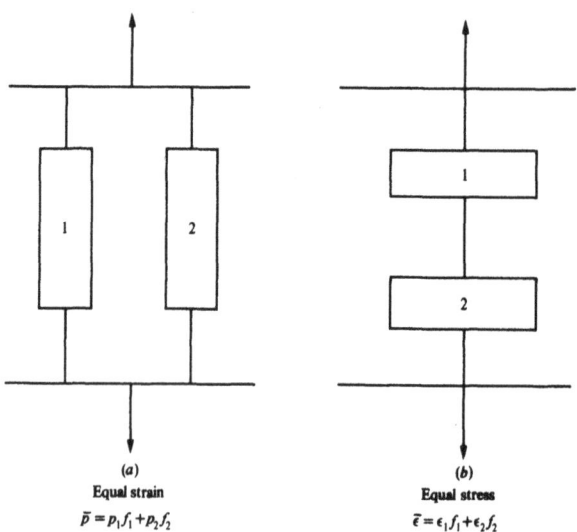

(a)
Equal strain
$\bar{p} = p_1 f_1 + p_2 f_2$

(b)
Equal stress
$\bar{\epsilon} = \epsilon_1 f_1 + \epsilon_2 f_2$

Fig. 2.34. Continuum models of the possible arrangements of the two constituents of an aggregate: (*a*) equal-strain model; (*b*) equal-stress model.

to the higher strain suffered by it. In both cases this mode of work-hardening continues until the yield stress of the harder constituent is reached. When the hard constituent yields, the work-hardening rate of the aggregate becomes virtually equal to that of the softer constituent but at a higher level of flow stress.

The real partition of stress between the constituents will obviously lie between the equal-stress and equal-strain cases. One empirical proposal is to partition the stresses according to the equal-stress model, and the strains according to the equal-strain model, so that if we define the volume fractions of the constituents as f_1 and f_2, we have

$$\bar{p} = f_1 p_1 + f_2 p_2, \tag{2.58}$$

$$\epsilon = f_1 \epsilon_1 + f_2 \epsilon_2. \tag{2.59}$$

This approach has been successfully used by Karlsson & Lindén (1975) to predict the flow stress of ferrite–pearlite structures in steels, in which

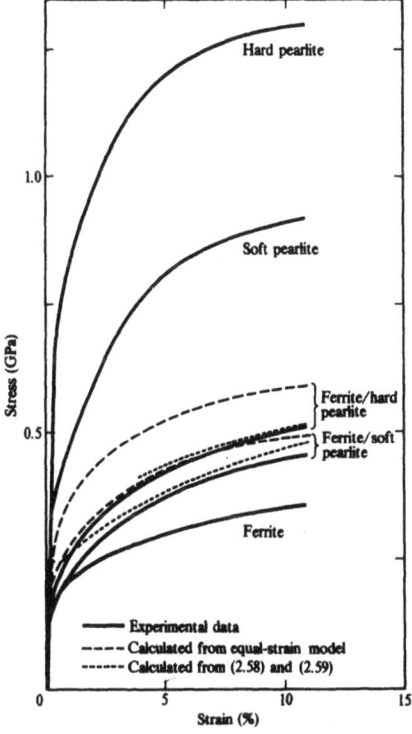

Fig. 2.35. Stress–strain behaviour of ferrite, pearlite, and the aggregate in a low-carbon steel. (Experimental and model results after Karlsson & Lindén, 1975.)

pearlite inclusions of differing hardness were produced by varying the cooling-rate after austenitization. The strain of the pearlite constituent was obtained from microhardness measurements, having first established the relationship between hardness and strain on deformed wholly pearlitic specimens. The ferrite strain could then be deduced from the pearlite strain, and the overall strain deduced by applying (2.59).

Combining the flow stresses for the constituents at their respective strains in the manner of (2.58), the flow stress for the aggregate material is obtained (fig. 2.35), which is seen to be in good agreement with experiment. In summary, therefore, it appears that the mixture law, stated in (2.58) and (2.59), allow one to predict with reasonable accuracy the stress–strain behaviour of coarse two-phase aggregates.

2.4   The achievement of strong particle-hardened structures

Having surveyed the theories of particle-hardening, and considered the extent to which the strength of particle-hardened alloys may be accounted for on the basis of these theories, we will conclude this chapter by reviewing how, in the light of our understanding, particle-hardened materials of high strength may be developed.

As we have seen, if the precipitates are deformable, the intrinsic properties of the precipitate are important, whereas the strength varies only weakly with precipitate size. For non-deformable precipitates, the strength is independent of the properties of the precipitate, but is strongly dependent on its dispersion and size. In both cases, the larger the volume fraction, the higher the strength. This situation is summarized in the diagram of fig. 2.36.

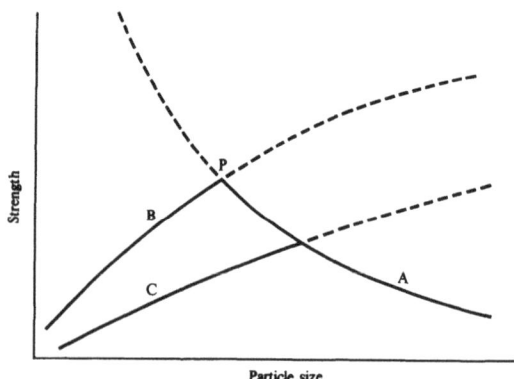

Fig. 2.36. Strength versus particle-size in particle-hardened material (see text). (After Nicholson, 1971.)

Curve B is for deformable precipitates, and shows the strength varying with particle size. Curve A shows the variation in strength with particle size for non-deformable particles with the same volume fraction as before. The transition from one type of behaviour to the other occurs at P, and this gives the maximum strength developed by the system.

An increase in volume fraction of precipitate raises both curves A and B to give a higher peak strength. An increase in the intrinsic strength of the precipitate will raise curve B only, and hence will increase the peak strength of the alloy, but this will occur at a smaller particle size. Therefore it is necessary to have a sufficiently prolific nucleation of the particles under these circumstances to ensure that the full volume fraction is precipitated by the time the average particle size reaches the critical value for peak strength. If the nucleation rate is inadequate, curve B will be lowered, say to curve C, which leads to a reduction in peak strength, and the displacement of the peak to larger particle sizes.

Unfortunately, there is a fundamental difficulty in achieving high volume fractions of a precipitate of high binding energy in alloys where the precipitation is a consequence of the temperature dependence of the solute solubility. This is because a high binding energy implies a small solubility, simply from the application of Hume-Rothery's rules. This effect is overcome in *steels* through the allotropic transformation in the base metal, which can be used as an alternative means of producing a high solute supersaturation in the low-temperature (ferrite) phase.

2.4.1   Maraging Steels

The most dramatic example of the use of this phase transformation to produce a high-strength, particle-hardened solid is the family of *maraging steels.* Initial studies of this process were carried out with Fe–18wt% Ni alloys of very low carbon content (0.03%). This solid solution was found to be martensitic on air-cooling from 860 °C, having a lath-like structure with a high density of mobile dislocations, which provided high ductility. A feature of this martensite is the wide thermal hysteresis between the austenite/martensite transformation on cooling and heating. The Fe–18wt% Ni alloy has an M_s on cooling of around 280 °C, but on heating the reversion temperature is about 600 °C, which allows high temperature (480 °C) ageing of the martensite matrix to be carried out without reversion to austenite.

The martensitic transformation occurs on air-cooling, from homogenization at 860 °C, and is usually carried out on rough-machined forged stock, the transformation being achieved throughout thick sections. The component can at this point be machined to its finished dimensions, since only extremely small-dimensional changes take place during the final hardening–ageing treatment. Strengthening during ageing is generally agreed to result from a very fine precipitation of Ni_3Mo and Ni_3Ti, although several other intermetallic phases have also been identified. These intermetallic phases are not coherent with the matrix, and so they nucleate preferentially at dislocations and grain boundaries. The high density of dislocations prior to ageing produced by the lattice invariant shear of the martensitic transformation thus ensures prolific nucleation sites for the dispersed phase. Yield strengths of around 1.8 GPa are attained in these steels together with elongations of 10% and reduction in area of 40 to 60%; the combination of these properties also leads to excellent fracture toughness and a range of high-technology uses.

2.4.2 Thermomechanical Treatment

Maraging steels owe their high strength to their high density of lattice defects plus the finely-dispersed phases in the microstructure. An alternative general method of producing such structures is that of *thermomechanical* treatment, which consists of a combination of ageing and deformation treatments in order to develop optimum properties. In ferrous alloys, the technique of *ausforming* is a good example of this type of process.

This treatment depends on the deformation of metastable austenite prior to quenching to martensite, and is schematically shown in fig. 2.37. By adding alloying elements (e.g. wt% Cr) which reduce diffusion rates and push back the austenite bay to very long times, a wide temperature range is made available for the existence of metastable austenite, which can then

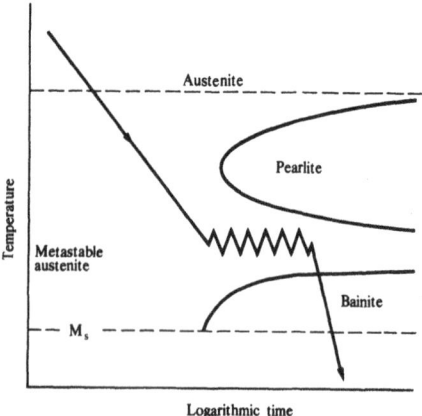

Fig. 2.37. Schematic time–temperature–deformation diagram of the ausforming process.

be extensively deformed. The austenite is then quenched and lightly tempered to produce a fine carbide dispersion, typically containing chromium, molybdenum and vanadium carbides. Again the enhanced properties arise from the combination of fine grain size, high defect density and finely-dispersed precipitates in the microstructure. The strength increase resulting from ausforming depends on the deformation temperature, the austenitizing temperature and the strain rate. As a general rule it can be stated that the yield strength and ultimate tensile stress (UTS) are increased by approximately 15 MPa for every 1% deformation.

To put these achievements in the strengthening of steel into perspective, the following table compares typical UTS values of various ferrous materials.

Ferrous material		UTS (MPa)
Single crystal of iron		26
Mild steel		340
Conventional alloy steel,	up to	2100
Maraged steel		2400
Ausformed steel		2940
Ideal strength of iron		8500

2.4.3   Dispersoids of high volume fraction

The limitations imposed by the equilibrium diagram to the achievement of high volume fractions of dispersoids may be bypassed in several ways. Historically, the earliest of these methods was that of powder metallurgy, and today the successful WC-Co 'hard metal' composite cutting materials are manufactured by this technique. Their hardness arises from the high volume fraction of the WC-phase, and their properties can be accounted for from a macroscopic 'law of mixtures' approach (§2.3).

Powder metallurgy provides a powerful means of dispersing phases of high thermodynamic stability (e.g. refractory oxides) in metals, although the approach of simply mixing the powdered metals and other oxides followed by compaction and sintering has had limited success. Firstly, uniform distribution of the oxides is a state which is extremely difficult, if not impossible, to achieve by such mixing methods, and secondly, the minimum interparticle spacing in the microstructure could never be less than that of the metal powder granule size itself. The successful powder-metallurgical methods include co-precipitation of finely-divided oxides followed by the selective reduction of one of them to the metal. Nickel-thoria composites ('TD-Nickel') are compacted and sintered following this type of technique, and more recently high-energy ball-milling ('mechanical alloying') has been employed to produce complex superalloys containing both oxide dispersoids and intermetallic (γ') phases which we will refer to in chapter 4.

In recent years another approach to the problems of the design of particle-hardened alloys has been initiated by the work of Duwez (1967) on 'splat cooling', which showed that non-equilibrium structures could be prepared by very rapid quenching of liquid alloys. Subsequent ageing of an appropriately highly supersaturated alloy produces a fine dispersion of precipitates. The splat-cooled granules have to be compacted and sintered to form a useful solid, but the scale of the dispersion is of course much finer than the granule size, in contrast to the particle spacings achievable by the direct mixing methods.

Finally, atoms of one metal can be injected into another by nuclear or diffusion techniques, thus avoiding restrictions on solid solubility. Ion implantation is an example of this principle, and precipitated microstructures of high volume fraction have been produced in this way. However, due to the high processing costs, no commercially viable material has yet been produced by this method.

3   Micromechanisms of fracture in two-phase materials

3.1   Introduction

Following our examination of the yield and flow behaviour of alloys containing precipitates, it is logical to proceed to the consideration of mechanisms of fracture of such materials. We will consider the effect of particles on the *nucleation* of fracture, and then upon the *propagation* of such nuclei until total failure occurs, discussing in turn the various modes of fracture, namely ductile rupture, crystallographic cleavage, and intergranular fracture. Finally, the role of particles in the micromechanisms of failure by fatigue will be considered.

Whereas the yielding and work-hardening of particle-hardened materials can be explained and even calculated in an approximate manner (chapter 2), this is not yet possible for the more complicated properties we have mentioned above. The processes of fracture are affected by the movement of relatively *large numbers* of dislocations, and it appears that the type of slip *distribution* occurring constitutes an important factor, and so we will consider this subject first.

3.2   Slip distribution in alloys containing particles

This topic, and its relevance to fracture behaviour has been reviewed by Hornbogen & Lütjering (1975). A given macroscopic amount of plastic strain can be distributed homogeneously or heterogeneously within a crystal. If one considers the case of slip on a single system, then fig. 3.1 indicates the principal geometrical possibilities. Fig. 3.1a defines a macroscopic shear strain, ϕ, and figs. 3.1b, c and d illustrate three ways in which

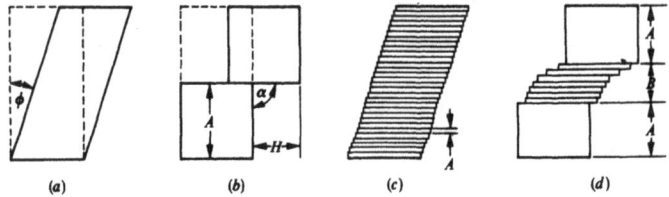

Fig. 3.1. Geometric possibilities for plastic deformation in a one-slip system. A is the slip step spacing; H is the height of the slip step; α is the angle between the slip plane and the crystal surface; B is the thickness of the slip band. (After Hornbogen & Zum Gahr, 1975.)

slip processes on a single system can provide this macroscopic strain. In fig. 3.1b, a slip step of height H has formed on a single plane which makes an angle α with the crystal surface. In fig. 3.1c the same macroscopic strain is effected by much smaller slip displacements upon many closely spaced planes, so that fig. 3.1b and fig. 3.1c provide extreme examples of heterogeneous and homogeneous slip distributions, respectively. In fig. 3.1d, the shear strain is provided by localized slip in a band of width B, the slip band spacing being A, and this corresponds to a situation intermediate between fig. 3.1b and fig. 3.1c in terms of slip heterogeneity.

If b is the Burgers vector on the slip system in question, and n is the number of glide dislocations in a particular slip plane, then

$$\phi = \frac{\Sigma H}{\Sigma A} = \frac{\Sigma n b \sin \alpha}{\Sigma A}. \tag{3.1}$$

Hornbogen & Zum Gahr (1975) have enumerated the factors that have an effect on slip distribution. Firstly, coarse slip is favoured by:

(1) low stacking-fault energy of solid solution;
(2) sheared precipitate particles;
(3) short-range order;
(4) radiation damage and holes;
(5) few slip systems operating;
(6) emissary dislocations;
(7) large grain size.

Secondly, fine slip is favoured by:

(1) high stacking-fault energy;
(2) bypassed particles;
(3) dislocation forest;
(4) dislocation climb;
(5) many slip systems operating;
(6) small grain size.

We will not discuss all these factors in detail here, but considering the role of stacking-fault energy, in a material where this energy is low, an increased stress is required to bring together dissociated dislocations to allow cross-slip. It is thus fairly obvious that heterogeneous slip will be favoured in such a metal. We have already established for second-phase particles that above a certain critical particle diameter (d_c), Orowan looping rather than particle shearing takes place (fig. 2.36), and that on further straining, dislocation debris accumulates round the particles. This leads to work-hardening of the slip plane upon which the dislocations have moved, thus favouring the operation of a different slip plane, i.e. slip tends to become *homogeneous*. This effect is illustrated in fig. 3.2. Fig. 3.2a is an electron micrograph of a crystal of a single-phase copper alloy

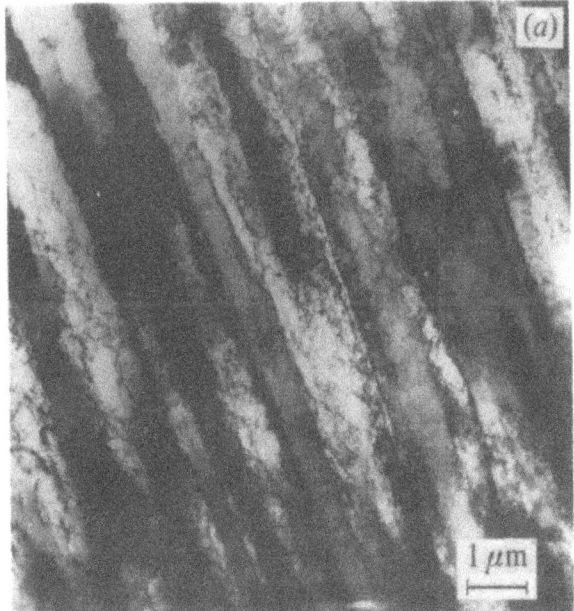

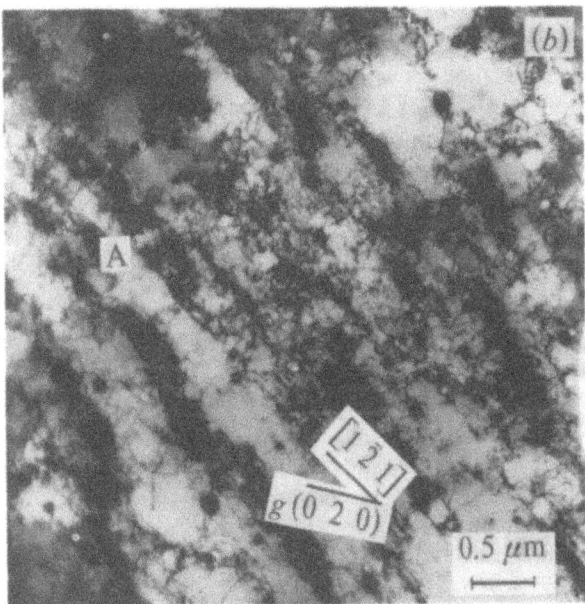

Fig. 3.2. (a) Single-phase Cu-0.7wt% Co crystal after 60% shear strain; ($\bar{1}$01) section. (b) Aged Cu-2.1wt% Co crystal after 64% shear strain; ($\bar{1}$01) section. (Courtesy of F. J. Humphreys.)

deformed into stage II of the tensile stress-strain curve, the specimen having been sectioned on the ($\bar{1}$01) plane, which is perpendicular to the primary (111) plane. Traces of the slip plane are well-defined, and are approximately 0.75 μm apart. The alternating light and dark contrast across the slip planes indicates lattice misorientations. These misorientations are found to be typically of the order of one degree. Fig. 3.2b shows the dislocation distribution in a two-phase crystal of Cu-2.1wt% Co alloy aged to form a dispersion of Co particles of average diameter 80 nm, and subjected to a shear strain comparable to that applied to the single-phase crystal. Comparison with fig. 3.2a shows that the dislocation distribution is profoundly changed: the primary slip planes can still be detected, but there is little contrast across them and no misorientations greater than 10′ were detected. This implies that, for equivalent stresses, slip distances and the number of dislocations per slip line are smaller in the crystal containing the unsheared particles than in single-phase alloys, so that the structure is less inhomogeneous, and lattice misorientations are smaller in magnitude.

This effect will, in general, be observed in crystals containing dispersions which are above the Orowan limit (d_c) in particle size (fig. 2.36). For dispersions of average diameter less than d_c, the particles are sheared at yield by an amount equal to the Burgers vector, b. On the slip plane, therefore, the effective cross-section of the particles is reduced after the passage of the first dislocation. Subsequent dislocations from the same source will further shear the particles and further soften the slip plane. This can be demonstrated more quantitatively by the following argument.

As shown in §2.1.1, the yield stress (τ) of a crystal containing a volume fraction f of particles of diameter d which are sheared by the dislocations at yield is usually given by an equation of the form

$$\tau = C f^{\frac{1}{2}} d^{\frac{1}{2}}, \tag{3.2}$$

where C is a constant whose value depends upon the hardening mechanism beind considered (see for example (2.14), (2.20) and (2.24)). If n dislocations of Burgers vector b shear a given particle, then the cross-section of the particle in the slip plane will be reduced by an amount nb. Assuming for simplicity that the particles are sheared across their diameter, then the stress for further shear becomes

$$\tau = C f^{\frac{1}{2}} (d - nb)^{\frac{1}{2}}$$
$$= C f^{\frac{1}{2}} d^{\frac{1}{2}} (1 - nb/d)^{\frac{1}{2}}. \tag{3.3}$$

The slip plane is thus work-*softened*, so that further slip will tend to concentrate on that plane, i.e. the slip will tend to be heterogeneous in such a crystal.

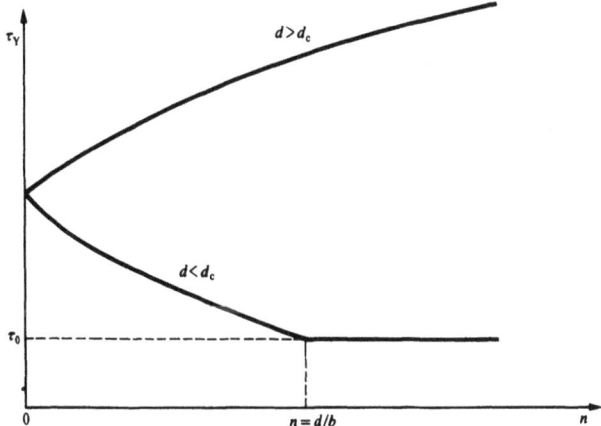

Fig. 3.3. Local critical shear stress, τ_Y ($= \tau_0 + \Delta\tau$), as a function of the number of dislocations, n, that have passed a slip plane. (After Hornbogen & Zum Gahr, 1975.)

The tendency to produce heterogeneous slip will be more pronounced the more τ decreases by the passage of one dislocation, so that the value of $d\tau/dn$ can be expected to be a property that characterizes the tendency for coarsening of slip due to work-softening. Differentiation of (3.3) gives

$$\frac{d\tau}{dn} = -\frac{bCf^{\frac{1}{2}}}{2d^{\frac{1}{2}}(1 - nb/d)^{\frac{1}{2}}}. \tag{3.4}$$

This indicates that the tendency for coarse slip is more pronounced the higher the volume fraction of the particles, the higher the value of C (which describes the specific hardness of the particle), and the smaller the particles are.

Fig. 3.3 summarizes the changes in local critical shear stress as a function of the number of dislocations that have passed a slip plane for the two situations discussed: that is, for particle sizes where $d > d_c$, when local work-hardening is observed leading to homogeneity of dislocation distribution; and also where $d < d_c$ and local work-softening according to (3.3) takes place.

Hornbogen & Lütjering have considered the changes in $d\tau/dn$, and hence the changes in the tendency for variations in slip distribution during an isothermal precipitation sequence in an age-hardening system. Their conclusions are shown schematically in fig. 3.4. In the single-phase solid solution the homogeneity of strain is dictated essentially by the stacking-fault energy of the material, as already discussed. When particles are nucleated, $d\tau/dn$ becomes increasingly negative (3.4), achieving a mini-mum value when the maximum volume fraction is precipitated at minimum

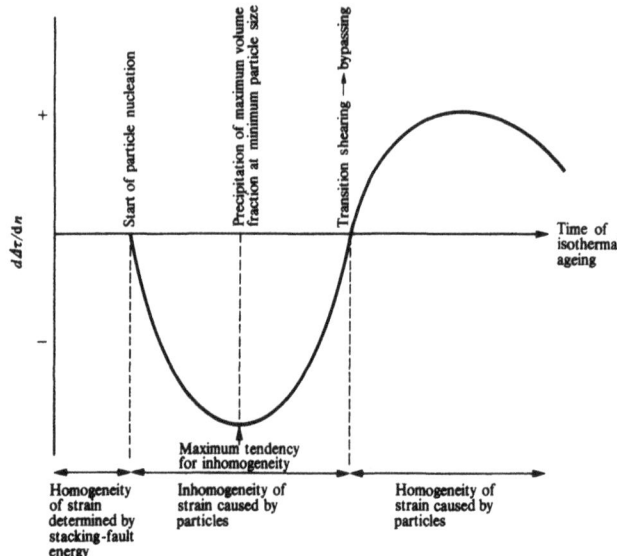

Fig. 3.4. Schematic illustration of the variation of $d\Delta\tau/dn$ during an isothermal ageing sequence of a precipitation-hardened alloy. (After Hornbogen & Lütjering, 1975.)

size. As the particles coarsen, $d\tau/dn$ becomes less negative, and when the particle size exceeds d_c dislocation bypassing occurs and the work-hardening of the slip plane leads to slip homogenization and thus positive values of $d\tau/dn$.

It is possible to demonstrate the above changes by studying the distribution of surface slip steps formed upon the polished surface of an alloy sample after ageing at different temperatures. Thus, Lütjering & Weissmann (1970) have compared the surface slip band distribution after 1% strain in an alloy of Ti–16.8wt% Al after ageing for 100 hours at 750 °C (fig. 3.5a) with that obtained after ageing for 100 hours at 850 °C (fig. 3.5b). The first ageing treatment gave rise only to a precipitation of coherent particles within the grains. The deformed surface shows relatively few slip steps, each with a large displacement. Ageing at the higher temperature led to the formation of partially coherent particles, which, when deformed, led to the operation of the dislocation circumvention mechanism. The surface is characterized in this case by the presence of a large number of small slip steps that are homogeneously distributed over the entire surface of the test piece. We will refer later to the differences in fracture behaviour in the two structures illustrated in figs. 3.5a and b; the main effect of an inhomogeneous distribution of plastic strain on the mechanical properties is that the fracture stress can be reached locally in the region of strain

Fig. 3.5. Slip-line distribution revealed by light microscopy in Ti–16.8wt%
Al: (a) annealed 100 hours at 750 °C then deformed (approximately 0%)
to fracture; (b) annealed 100 hours at 850 °C then deformed 1%. (After
Lütjering & Weissman, 1970.)

concentration, even if the overall (macroscopic) plastic deformation is still
relatively small. We will now consider the various possible fracture
processes in turn, and relate, where appropriate, the effect of slip distribution upon the various mechanisms.

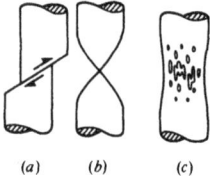

(a)   (b)   (c)

Fig. 3.6. The mechanisms by which a cylindrical tensile specimen may fail by ductile failure: (a) idealized simple shear; (b) idealized simple necking; and (c) the void growth process.

3.3   Ductile rupture

Failure in ductile solids arises from the concentration of plastic deformation into a localized region. An idealized model of the process is illustrated in fig. 3.6a in which, under an applied tensile strain, simple shear occurs along a single slip line. This concentration of plastic strain cannot take place so long as the material is still work-hardening strongly. If two orthogonal slip lines are operative, then the process illustrated in the idealized model of fig. 3.6b will occur, giving rise to simple necking. The processes illustrated in figs. 3.6a and b will produce knife-edge or point failures, but in practice most ductile failures give much less than total reduction in area. This is because ductile rupture usually proceeds simultaneously from a number of holes or voids inside the material.

The most fundamental discovery concerning the mechanism of ductile rupture in materials containing second-phase particles was that it is initiated by the generation of voids at the particles – either by decohesion of the particle/matrix interface or by cracking of the second-phase particles themselves. These cavities grow as the material is strained further, and ultimately the cavities coalesce by an *internal necking* mechanism to give the dimpled fracture surface characteristic of ductile failure. This void growth process is illustrated diagrammatically in fig. 3.6c, and we will consider separately the nucleation and the growth of such internal voids during the ductile rupture of particle-hardened materials.

3.3.1   Nucleation of voids at particles

Although the processes of void growth and interlinkage to final fracture are not yet fully understood, they are in a far better level of development than the initial processes which lead to the nucleation of holes from second-phase particles. It has been generally observed that while inclusions with large aspect ratio may undergo multiple internal fracturing, equiaxed inclusions almost always nucleate holes by interfacial separation. The problem of non-equiaxed particles, whose behaviour depends on their shape and orientation in addition to their size and spacing, is a complex one, and will not be discussed here. We will thus confine our attention to the process of void nucleation at equiaxed particles, regarding them as rigid and plastically non-deformable.

The theories that have been developed for void nucleation. can be grouped into three categories: (i) energy criteria; (ii) local strain criteria; and (iii) local stress criteria. We will consider these briefly in turn. A more detailed review has recently been published by Goods and Brown (1979).

Critical energy criteria

This model assumes that, as the matrix flows, the regions within and around the particles must increase in elastic energy, and that at a certain strain it is energetically favourable for this locally concentrated strain-energy to be released by decohesion or fracture of the particle. Clearly, this model is incorrect, because it assumes that the whole complex array of loops of dislocations surrounding a particle collapses into the interface when the cavity forms; in practice most of the loops must be immobilized by intersections with each other.

If we consider simply the elastic energy released by a (spherical) particle, and equate this to the energy of the surface formed by cavitation, we may write

$$\frac{2}{3}\pi r^3 G \gamma_n^2 = 4\pi r^2 \sigma$$
$$\gamma_n = (6\sigma/rG)^{\frac{1}{2}}, \tag{3.5}$$

where r is the particle radius, and σ is the specific surface energy of the crack and may be written as

$$\sigma = \sigma_m + \sigma_p - \sigma_i,$$

where σ_m is the surface energy of the matrix, σ_p that of the particle and σ_i that of the particle/matrix interface.

Equation (3.5) thus predicts that the nucleation strain (γ_n) is dependent upon the particle size, and that decohesion would first occur at the largest particles. Tanaka, Mori & Nakamura (1970) have calculated that the appropriate energy considerations are satisfied for all but the smallest particles (less than 25 nm) almost as soon as yield occurs. Since, in many instances, inclusions of more than one hundred times this size have been observed to remain cohering to the matrix after strains of more than one hundred times the yield strain have been applied, it must be concluded that the energy requirement is only a *necessary* one, i.e. separation requires that the interfacial strength is reached at some local points.

Critical strain criterion

McClintock (1968) has suggested that cavity formation at interfaces may obey a critical local strain criterion, or alternatively a critical interfacial shearing strain and an interfacial normal stress. The nature of this combined criterion has not been clarified, but McClintock gives an extensive elastic/plastic continuum analysis of stress distributions around non-deforming particles in an ideal plastic matrix, and shows that large strain concentrations can develop around the particles.

Critical stress criterion

Approaches involving critical local stresses and strains have been advanced to try to forecast a *sufficient* (as well as a *necessary*) condition for cavitation. The concept of a critical local stress for void generation can be invoked from the discussion of relaxation mechanisms at undeformable particles in a plastic matrix outlined in §2.2.1. As indicated in fig. 3.7, when the matrix is sheared compressive stresses develop in two quadrants adjacent to the particle, and tensile stresses in the other two. Arrays of prismatic dislocation loops are indicated in each quadrant. Ashby (1966) calculated the stress on the particle/matrix interface arising from a stack of such loops, and assumed that void nucleation would occur (in a 'tensile' quadrant) when this stress in the boundary reaches some assumed fracture stress of the boundary.

If there are n loops in a stack of length k, the average spacing of the loops in the stack is thus k/n, which is also the distance of the nearest loop from the particle. Close to a loop the stresses look like those near a

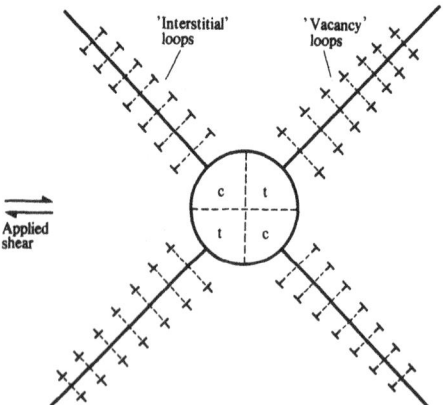

Fig. 3.7. Arrays of secondary prismatic loops at a hard particle in a plastic matrix subjected to the indicated applied shear. Compressive (c) and tensile (t) regions of the particle/matrix interface are indicated.

straight dislocation, and fall off inversely with distance, so we can consider the tensile stress on the particle/matrix interface to be due to the nearest loop. The tensile stress on the particles thus increases as n/k; but $n = 4 r\gamma/b$, from (2.50), so it increases as $4 r\gamma/k$ b. Thus, the stress increases approximately linearly with strain and particle diameter.

In a given specimen, large particles of a given phase will therefore cavitate before small ones, and a specimen containing small particles can be strained further without cavitation than one containing large particles.

If the inclusions are large ($\gg 1$ μm), and if the material containing them does not show coarse slip, it is adequate to treat them as if they were embedded in a plastic *continuum*. Argon, Im & Safoglu (1975) have made approximate continuum analyses for extreme idealizations of matrix behaviour and showed that in pure shear loading, the maximum local tensile stress at a spherical inclusion is then between 1.5 and 2 times the remote applied stress. An interesting feature of this treatment is that the particle-size dependence of γ_n can be associated with a non-uniformity in the distribution of the inclusions – the interfacial tensile stresses increasing with the *local* volume fraction. At very small volume fractions of second phase, the inclusions do not interact for very substantial amounts of plastic strain, and in this regime the interfacial stress is calculated to be independent of inclusion size. Argon, Im & Safoglu (1975) conclude that some of the many reported instances of inclusion-size effect in cavity formation can be explained in their model by variation of the volume fraction of second phase from point to point (i.e., in practice, there will be a non-uniform distribution of inclusions).

Ashby (1977) has tabulated some values of experimentally observed nucleation strains for cavities at second-phase particles in specimens subjected to simple tension, and these are given in table 3.1. Since the techniques used to determine these strains (expressed as a tensile logarithmic strain, ϵ_n) differed widely, the results are not all mutually consistent, but they will serve to indicate the order of magnitude of ϵ_n in practice.

3.3.2   Growth and coalescence of voids

Having nucleated voids of particles, the holes then grow as the applied tensile strain increases until they coalesce to give a fracture path. A number of the early theoretical models for this process have been reviewed by Knott (1973), but a recent simple theory due to Brown & Embury (1973) leads to a good description of the experimental observations.

Brown & Embury's treatment accounts semi-quantitatively for much of the data, and is amenable to comparison with a variety of microstructures.

Table 3.1. *Approximate nucleation strains* (ϵ_n) *at room temperature in simple tension for voids at second-phase particles.* (After Ashby, 1977.)

Material	Inclusions	Matrix	ϵ_n	Reference
Internally oxidized	SiO$_2$	Cu	0.1–0.2	Palmer and Smith (1968)
copper alloys	SiO$_2$	Cu	0.1–0.2	Atkinson (1973)
	Al$_2$O$_2$	Cu	0.1–0.2	Gould and Humphreys (1973)
	BeO	Cu	0.2–0.4	Gould and Humphreys (1973)
	SiO$_2$	-brass	≈ 0	Humphreys and Stewart (1972)
Spheroidized carbon	Fe$_3$C	ferrite	0.4	Brown and Embury (1973)
steel	Fe$_3$C	ferrite	0.3–1.0	Inoue and Kinsohita (1973)
Swedish iron	oxides	ferrite	<0.2	Hancock (1976)
Cu–0.6wt% Cr	Cu–Cr	copper	≈ 1.0	Argon and Im (1975)
1045 spheroidized steel	Fe$_3$C (12.5%)	ferrite	0.6–0.7	Argon and Im (1975)
Maraging steel	TiC (1.1%)	ferrite	≈ 0.8	Argon and Im (1975)
HY 130	carbides	ferrite	≈ 0.01	Hancock and MacKenzie (1976)

After the voids are nucleated (fig. 3.8*a*), it is assumed that they grow by plastic extension, although the ductile matrix between neighbouring voids cannot suffer very large local deformations because it is under plastic constraint by the surrounding material. Because of its stress-concentrating effect, a spherical void will elongate initially at a rate of about twice that of the specimen itself. As it extends and becomes ellipsoidal, it grows more slowly until, when very elongated, it extends at the same rate as the specimen itself.

When the voids have been extended so that the spacing of neighbouring voids becomes equal to their length (fig. 3.8*b*), a plastic slip line field can be constructed between the voids, as indicated in the diagram, so that the plastic constraint preventing local deformation is removed. Brown & Embury assume that once the geometry illustrated in fig. 3.8*b* is achieved, any further plastic flow is localized on one slip line, and ductile fracture ensues immediately (fig. 3.8*c*) – it is assumed that the contribution of this final linking strain to the total reduction in area will be very small.

It should perhaps be emphasized that, in a tensile test, macroscopic necking will normally have occurred well before the geometrical condition represented in fig. 3.8*b* is met, as the former is defined by the usual Considère condition. The model fails to take into account any influence such a neck might have on the nucleation and growth of holes, and this must be regarded as a shortcoming of the theory.

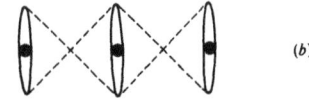

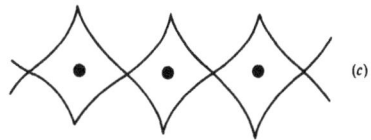

Fig. 3.8. The growth, (*a*) and (*b*), and coalescence (*c*) of voids at particles during ductile failure according to the theory of Brown & Embury (1973).

The final fracture surface, corresponding to the result of the processes illustrated in fig. 3.8 will consist of an array of dimples, each dimple containing a particle. This is confirmed experimentally, as illustrated in fig. 3.9. (Some particles will be attached to the mating fracture surface.)

Brown & Embury's model enables an estimate to be made of the tensile strain e_g required to cause the voids to grow to the unstable configuration illustrated in fig. 3.8*b*. If the diameter of the particles is $2r$, then $2r$ is also the size of the void when it is nucleated. If the particle spacing is λ, then a void length λ is required to enable the slip line field to be constructed (fig. 3.8*b*). If the voids elongate at a rate which, on average, is faster by a factor C than the rate of extension of the specimen itself ($1 < C < 2$), then we may write the true strain to coalescence as

$$e_g = \frac{1}{C} \ln (1 + e_g), \tag{3.6}$$

where $e_g = (\lambda - 2r)/2r$,

i.e. $\lambda = 2r(1 + e_g)$. \hfill (3.7)

Since the fracture path is approximately a plane, λ is identified with the mean two-dimensional or planar spacing of the particles. As discussed in chapter 1, we may write

$$\lambda = N_S^{-\frac{1}{2}} - 2r_S,$$

where $r_S = \sqrt{\frac{2}{3}}\, r$, see (1.31). If the volume fraction of the particles is f,

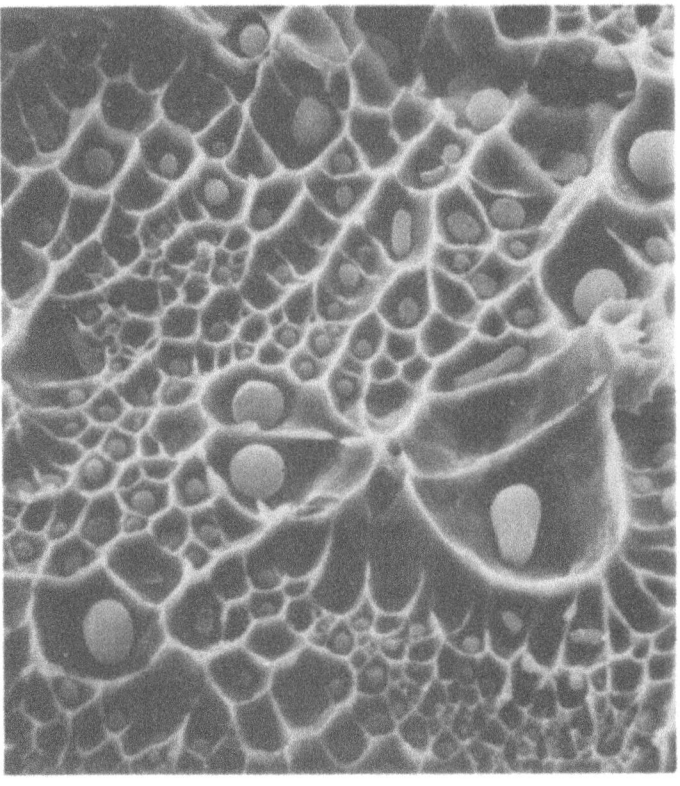

10 μm

Fig. 3.9. Ductile fracture surface of low-carbon steel containing sulphide particles, showing the relationship between the particle size and the ductile cusp size. (Courtesy of T. R. Gladman.)

then $N_S = f/(\pi r_S^2)$, and appropriate substitution in (3.7) gives

$$2r(1 + e_g) = r\left(\sqrt{(2\pi/3f)} - \sqrt{\tfrac{8}{3}}\right). \tag{3.8}$$

Equation (3.8) indicates that the growth strain becomes zero above a critical volume fraction of $f = 0.159$, which is when a slip line can be drawn between neighbouring *particles* even in the absence of voids. For volume fractions greater than 0.159, therefore, the fracture strain is controlled entirely by the *nucleation* strain (e_n). Substituting from (3.8) for $(1 + e_g)$ in (3.6) to give the true strain, we have

$$\epsilon_g = \frac{1}{C}\ln\left(\sqrt{(\pi/6f)} - \sqrt{\tfrac{2}{3}}\right),$$

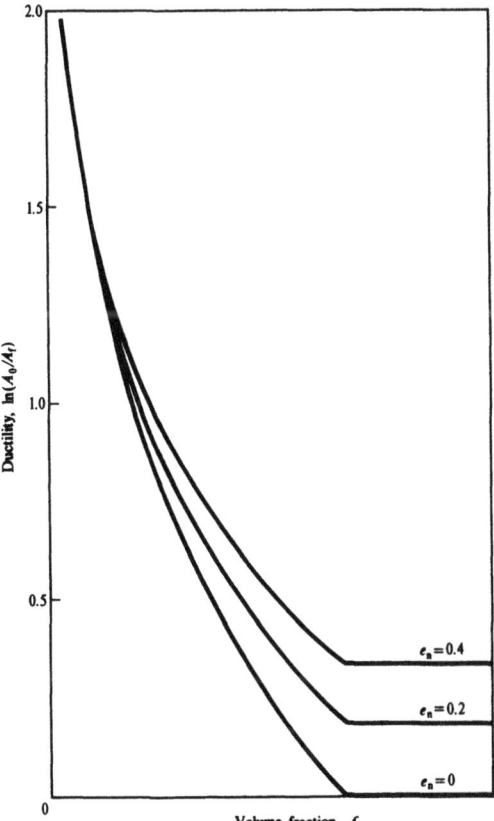

Fig. 3.10. (*a*) The relationship between ductility, ln (A_0/A_f), and the volume fraction, f, according to (3.9) for three assumed values of nucleation strain, namely 0, 0.2 and 0.4.

and we may write the total strain to failure as

$$
\begin{aligned}
\epsilon_f &= \epsilon_n + \epsilon_g \\
&= \frac{1}{C} \ln (1 + e_g + e_n) \\
\text{that is} \quad \epsilon_f &= \frac{1}{C} \ln (\sqrt{(\pi/6f)} - \sqrt{\tfrac{2}{3}} + e_n) \\
&= \ln (A_0/A_f)
\end{aligned}
\Biggr\} ,
\tag{3.9}
$$

where A_0 and A_f are the initial and final cross sectional areas of the specimen respectively. Taking the value of C as 2, fig. 3.10*a* illustrates

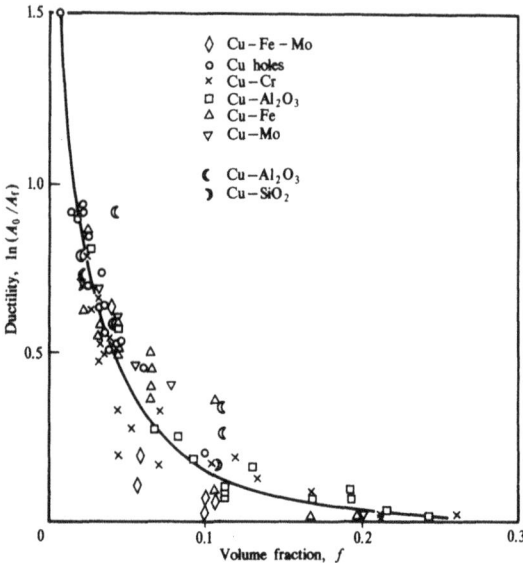

Fig. 3.10 (*b*) A combined plot of ductility of several copper dispersion alloys versus volume fraction (after Edelson & Baldwin, 1962).

the relationship between ductility, $\ln(A_0/A_f)$, and volume fraction, f, expressed in (3.9) for three assumed values of e_n, namely 0, 0.2 and 0.4. Fig. 3.10*b* illustrates the data of Edelson & Baldwin (1962) for some dispersion-strengthened copper-based alloys which appear to be in reasonable accord with the model. In technological materials, the effect of small strengthening particles and macroscopic inclusions is not additive in defining f in (3.9), and since decohesion of the large particles may occur at very low strains, the large voids so formed may serve to limit the volume of material which needs to be strained to ϵ_f.

Another shortcoming of Brown & Embury's model is that it fails to include the effect of the formation of zones of concentrated shear. For example, in a material which exhibits inhomogeneous slip distribution, a ductile crack may (depending on the degree of slip inhomogeneity) propagate along slip bands in the matrix. Localized shear of this kind can cause holes to coalesce at relatively small macroscopic strain, giving failure by 'void-sheeting'. This effect is observed in Al–Cu alloys in the underaged condition, i.e. when containing a dispersion of GP zones: deformation occurs in intense bands, and cracks are nucleated in these bands to produce a characteristic shear fracture.

3.4   Transgranular cleavage

Cottrell (1958) proposed a dislocation mechanism for cleavage fracture which recognized the importance of a microstructure in determining the stress for the plastic deformation mechanism that precedes crack formation. His well-known criterion for brittle fracture may be written as

$$\tau_Y \, k_Y \, d^{\frac{1}{2}} \geqslant 2\beta \, G \, \sigma, \qquad (3.10)$$

where β is a factor given by the ratio of the maximum shear stress to the maximum tensile stress in the specimen. G is the shear modulus and σ is the surface energy per unit area of the cleavage crack. The terms on the left-hand side of the equation are related through the Hall-Petch equation

$$\tau_Y = \tau_i + k_Y d^{\frac{1}{2}}, \qquad (3.11)$$

which relates the yield stress in shear (τ_Y) to the grain size (d). The term k_Y relates to the process by which mobile dislocations are produced in the unyielded grains.

Cottrell's model, therefore, emphasizes the role of tensile stress and explains the effects of grain size and yielding parameters on fracture. Initially (3.10) appeared to provide a complete basis for the microstructural design of materials with improved resistence to cleavage. The model does not, however, consider the effect of microstructural inhomogeneities on the initiation and propagation of cleavage cracks, and is therefore not applicable to some steels where cleavage cracks are initiated by the fracture of grain-boundary carbides.

This was demonstrated by McMahon & Cohen (1965) who tested tensile specimens of identical yield and flow properties, but containing carbides of differing sizes dispersed at the grain boundaries. One steel had discontinuous carbide films 3 to 10 μm thick while the other had grain-boundary carbide particles that were only 1 to 3 μm thick. It was shown that the fracture properties were quite dissimilar, in contradiction of the Cottrell theory. The ductility transition for the steel with the thick carbide was -90 °C and for the other steel it was -160 °C. Since for both steels there were numerous cracks within carbide particles that had not propagated to give ferrite cleavage cracks, the critical event in ferrite cleavage crack formation appears to be the *growth* of a crack from within a particle into the surrounding matrix.

3.4.1   Nucleation of cleavage from grain-boundary particles

These observations led to an alternative model for growth-controlled cleavage fracture which incorporated the effect of carbide particles. This

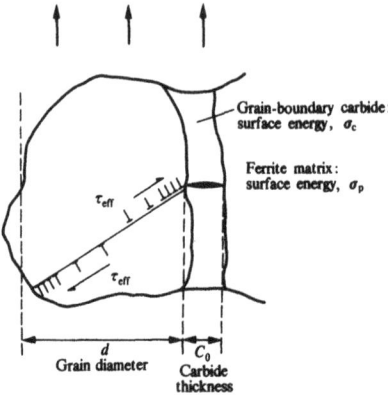

Fig. 3.11. Model for cleavage fracture due to Smith (1966).

model, due to Smith (1966), proposes that nucleation of a crack within a grain-boundary carbide particle occurs as a direct consequence of the high local stresses which are generated by inhomogeneous plastic deformation processes in the form of a pile-up of dislocations in the surrounding ferrite (fig. 3.11) of grain size d. It may be deduced that a carbide of thickness C_0 will be cracked by the pile-up if

$$\tau_Y - \tau_i = \tau_{\text{eff}} \geqslant (4 E\, \sigma_c / \pi d\, (1 - \nu^2)^{\frac{1}{2}}, \tag{3.12}$$

where σ_c is the surface energy of the carbide particle, and E its Young's modulus. The change in energy required to spread the crack from the carbide into the ferrite matrix may be examined, in a manner similar to that for the Cottrell model, to obtain a failure criterion (for growth-controlled failure under a critical fracture stress in tension, p), given by

$$(C_0/d)p_{\text{f}}^2 + \tau_{\text{eff}}^2\, [\tfrac{1}{2} + (2\tau_i/\pi\tau_{\text{eff}})(C_0/d)^{\frac{1}{2}}]^2 \geqslant 4 E\, \sigma_p / \pi d\, (1 - \nu^2), \tag{3.13}$$

where σ_p is a surface energy or plastic work term for cleavage of ferrite. If the substitution is made in (3.13) of $k_Y d^{-\frac{1}{2}}$ for τ_{eff} (3.11), it is found that the grain-size parameters cancel, leaving a relationship whose only microstructural parameter is carbide width, C_0.

In view of the commonly observed grain-size dependence of cleavage fracture stress in steel, Knott (1977) suggests that a possible reason for the apparent anomaly is that the development of microstructure by cooling from the austenitizing temperature is such as to maintain the ratio of C_0/d constant, since both C_0 and d depend on similar diffusion phenomena. Knott reports work by Curry confirming a strong positive relationship between the two parameters (although it is not quite linear).

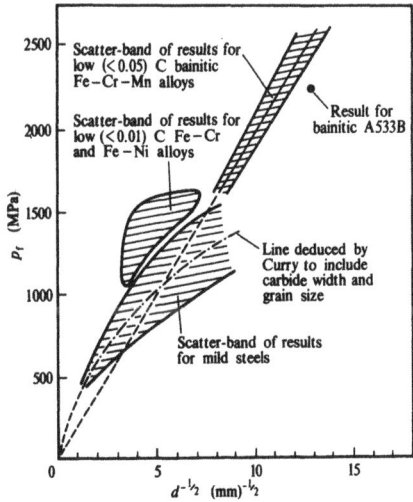

Fig. 3.12. Experimental values of local fracture stress, p_f, plotted versus the reciprocal square root of the grain size, $d^{-\frac{1}{2}}$, for a number of ferrous materials. (After Knott, 1977.)

By using these data and (3.13) he constructed a predicted failure line (shown schematically in fig. 3.12) whose precise position depends on σ_p (taken as 14 J m^{-2}) and the percentile of the carbide thickness distribution chosen to represent equally critical crack nuclei.

Knott further quotes an extension of Smith's model by Curry to cover low- and medium-carbon steels heat-treated to produce spheroidal carbides. Here the dislocation contribution to propagation is ignored because pile-ups do not occur but dislocation cells form instead. The crack nucleus is taken as penny-shaped, since it forms in a spheroidal particle, so that the expression for p_f is

$$p_f = (\pi E \, \sigma_p / 2 \, C_0)^{\frac{1}{2}}. \tag{3.14}$$

3.4.2   The effect of particles upon the propagation of cleavage

The choice of cleavage plane

The influence of fine-scale precipitate dispersions upon cleavage crack propagation in steels is not well understood. Averbach (1974) states that the cleavage plane in martensite is influenced by the habit plane of carbides precipitated during tempering. Tempering at 260 °C gave ϵ-carbide on $\{100\}_\alpha$ planes, and considerable cleavage was also noted on these planes in fractured thin-foil transmission electron microscopy (TEM) specimens. Tempering at 340 °C and 480 °C gave cementite on $\{110\}_\alpha$ planes, which then appeared as an additional cleavage plane.

Benson & Edmonds (1977) have recently examined cleavage crack propagation in low-alloy ferritic steels containing aligned fine-scale carbide dispersions resulting from periodic precipitation at the austenite/ferrite interface during austenite decomposition. No firm evidence could be found to suggest that the aligned sheets of precipitates influenced the cleavage plane, which was shown to be consistent with $\{100\}_{\alpha}$ at $-196\,^{\circ}\mathrm{C}$ for the steels employed.

Quenched and tempered alloy steel

Kotilainen & Törrönen (1977) have examined and evaluated the fracture behaviour of a Cr–Mo–V steel which had been quenched and tempered at various temperatures to a microstructure of tempered granular bainite. The microstructural element which controlled the yield strength was a fine dispersion of vanadium-rich MC-type carbides. The yield strength obeyed an Orowan relationship with regard to this phase. The properties of this steel are contrasted with 'Petch-materials', the *grain size* of which controls both the yield and cleavage fracture strength.

The fracture stress was determined as a function of temperature by means of instrumented impact tests, and it was compared with that calculated from the Griffith equation (3.12). Good agreement was obtained if d in (3.12) was equated with the bainite lath packet width (not the individual lath width, or the carbide particle spacing), which suggests that high-angle boundaries are strong enough obstacles to stop microcracks, while small carbides or low-angle lath boundaries are not.

Kotilainen & Törrönen conclude that crack nucleation and initial growth occurs with difficulty because the fine dispersion of carbides restricts the necessary dislocation motion to form a stress concentration. The good toughness which the steel exhibits is thus due to this effect and also to the small width (1.33 μm) of the lath packets, whose boundaries arrest growing microcracks.

The propagation of these microcracks is, in fact, easier than their nucleation. The slowing and arresting of a propagating microcrack depends on the possibility of some plastic deformation at the crack tip absorbing energy. The dislocations are locked by the MC dispersion and thus the crack can propagate without losing energy. The carbide dispersion is, therefore, beneficial in raising the stress required for crack nucleation, but it is ineffective against crack propagation since the dispersion determines the ease of plastic deformation. This implies that the ductile/brittle transition temperature in Charpy V impact tests should be linearly related to the inverse of the carbide spacing (as is the yield stress), and this is shown to be so in fig. 3.13.

This type of steel thus differs in its fracture behaviour from that of steels whose strength and toughness has been achieved by grain refinement.

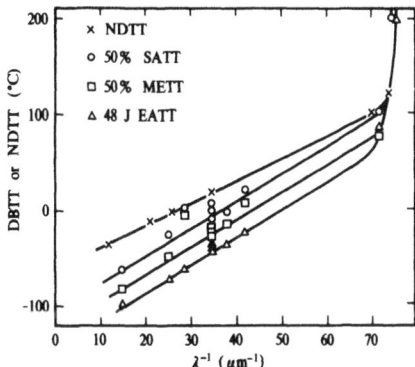

Fig. 3.13. The Charpy V ductile/brittle temperature (DBTT) and Pellini nil-ductility transition temperature (NDTT) as a function of planar spacing, λ, of MC carbides in a tempered Cr–Mo–V pressure-vessel steel. The DBTT has been assessed in three ways: 50% SATT, the temperature corresponding to the 50% shear-lip area; 50% METT, the temperature corresponding to an energy that is the mean of the difference between maximum and minimum energies; 48J EATT, the temperature corresponding to an energy of 48 J. (After Kotilainen and Törrönen, 1977.)

Effects in brittle solids

The more general problem of the effect upon crack propagation of hard particles dispersed in a brittle solid has been considered by Evans (1972) in an attempt to account for the increase in fracture strength brought about by such dispersions. The increase in strength may be due to a line tension effect as the crack bows between the obstacles in the same way that a dislocation bows between obstacles in its slip plane (fig. 3.14). The fracture resistance of the intersected obstacles is assumed to be large compared

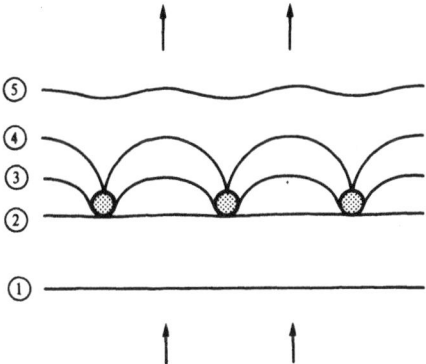

Fig. 3.14. Sequence showing successive positions of a crack front upon interaction with an array of second-phase obstacles.

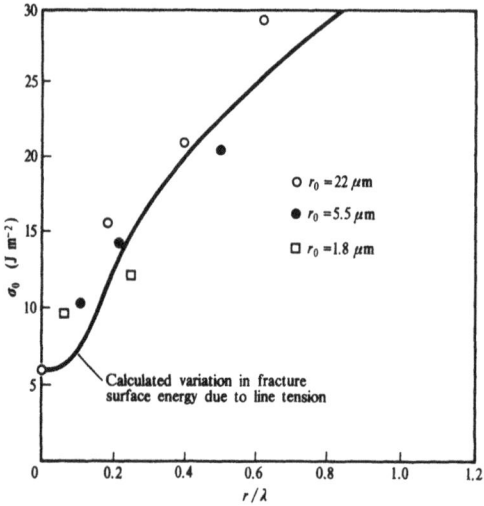

Fig. 3.15. The fracture energy, σ_0, for a borosilicate glass/alumina system compared with calculated line tension contribution for spherical impenetrable obstacles. (After Lawn & Wilshaw, 1975.)

with that of the matrix, so that the crack is effectively pinned at points along its front. Under these conditions the Griffith equation for the fracture stress p_f may be written as

$$p_f = (2 E \sigma_0 / \pi C)^{\frac{1}{2}}, \tag{3.15}$$

where C is the crack length and

$$\sigma_i = \sigma_0 + \sigma_a + T/2\lambda, \tag{3.16}$$

where σ_0 is the fracture surface energy of the matrix material, σ_a is a term included to account for any enhanced plastic deformation that occurs in the matrix adjacent to the obstacle, T is the line tension of the crack front and λ is the obstacle spacing. From a calculation of the magnitude of the line tension of a propagating crack in the maximum stress position indicated in fig. 3.14, the crack extension force is predicted to increase with the ratio of particle diameter to interparticle spacing. Fig. 3.15 illustrates the calculated change in fracture surface energy with this parameter, which may be compared with experimental data obtained by Lange (1971) for dispersions of alumina particles in glass.

3.5   Intergranular fracture

The phenomenon of brittle intergranular cleavage involves the fast propagation of cracks along grain boundaries which have been weakened either by the accumulation of minor impurity elements or by the presence of a continuous film of a brittle phase. It is also possible for fracture to assume an intergranular path, but to progress by the linkage of voids formed around second-phase particles. This effect will be found in situations where the density of weakly-bonded particles within the grains is low, but that on the boundaries is high. We will consider each of these intergranular modes in turn.

3.5.1   Brittle intergranular cleavage

In precipitation-hardening systems, the importance of slip *distribution* may be critical in determining the macroscopic ductility. This distribution, as discussed in §3.1, is dependent upon the state of precipitation, and as an example of this effect we will return to the age-hardened Ti-Al alloy investigated by Lütjering & Weissmann (1970), illustrated in figs. 3.5a and b. When this alloy was aged to produce a dispersion of the coherent Ti_3Al-phase, dislocation cutting of the particles occurred at yield, giving rise to sharp, widely-spaced slip bands (fig. 3.5a). The corresponding stress–strain curves did not show any elongation at all. By contrast, the same alloy, when aged at 850 °C for 100 hours became more ductile (tensile fracture strain 3 to 5%), and the coarse slip mode is replaced by fine, more homogeneous slip (fig. 3.5b) associated with a dislocation bypass process.

This example illustrates a general concept: that alloys containing a homogeneous distribution of ordered coherent particles, which upon deformation are sheared by the moving dislocations, are usually very brittle. The elongation to fracture of such alloys can, however, be improved by an ageing treatment which aims to make the interparticle distance so large that the moving dislocations have to bypass the particles instead of shearing them. Under conditions of particle-cutting and the development of intense, narrow slip bands, high-stress concentrations will form at the head of dislocation pile-ups at grain boundaries. The local stress p^* at the head of a pile-up of n dislocations against a grain boundary under an external stress of p_a is given by

$$p^* = n\,p_a. \tag{3.17}$$

p^* promotes the crack nucleation at the boundary (fig. 3.15) and subsequent fracture at negligible macroscopic strain. The critical value of p^* that leads to crack nucleation and fracture undoubtedly depends on the strength and deformability of the grain boundary. If the slip is homogenized by changing the dispersion of the precipitate, n in (3.17) is reduced, so that in order to develop the critical grain-boundary stress for fracture, a greater applied strain is necessary – hence the observed increase in ductility in this circumstance.

3.5.2   Intergranular fibrous fracture

A situation where a high density of weakly-bonded particles exist on the grain boundaries is in an overheated (low-sulphur) steel, which fails round the prior austenite grain boundaries. At high temperatures, the majority of the sulphide particles in such a steel will be taken into solution. On cooling at an appropriate rate, a fine distribution of sulphide particles forms preferentially upon the austenite grain boundaries, which provides a low-energy path for intergranular fibrous fracture.

Somewhat more complex examples of the phenomenon may be met in precipitation-hardening systems, and we will examine these in more detail in view of their technological importance.

Nucleation

Here again, the importance of slip *distribution* within the grains themselves is important. We have seen how by overageing an age-hardening alloy, slip homogenization develops owing to the change in the dislocation/particle interaction process. Slip homogenization produced by these means is thus inevitably accompanied by a decrease in yield stress. In many commercial aluminium alloys, however, slip homogenization arises through the presence, in addition to the coherent phase which gives rise to the high yield, of a coarser intermetallic dispersoid. This is because, as mentioned in §1.1.3, a transition metal addition (e.g. Cr, Zr or Mn) is made to such alloys which then form precipitates of noncoherent intermetallic compounds whose size range is typically 0.1 to 1 μm (e.g. see fig. 1.1). These dispersions have the effect of maintaining a fine grain size in the alloy during processing, and may also affect the deformation mode of the alloy. This effect is illustrated in figs. 3.16*a* and *b* showing dislocation distributions in plastically strained Al-0.6wt%Mg-1wt% Si alloys in the peak-aged condition. The specimen shown in fig. 3.16*a* is a simple ternary alloy, and the coherent Mg_2Si-phase is sheared on yielding so that intense bands of dislocations are evident. Intense surface slip bands also form, as in the Ti alloy of fig. 3.5*a*. The specimen of fig. 3.16*b* is of similar composition to that in fig. 3.16*a*, except that it contains 0.5wt% Mn which is present as a dispersion (fig. 1.1) of incoherent particles of the α-$Al_{12}Mn_3Si$-phase. This specimen has also been aged to (the same) peak hardness, so that both the coherent hardening phase and the incoherent dispersoid are present. On plastic deformation the dislocation distribution in this specimen is much more homogeneous than in the ternary alloy.

The Mn-free alloy exhibits about 1% elongation to fracture in a tensile test, the fracture being intergranular (fig. 3.17*a*). At higher magnification (fig. 3.17*b*) the fracture surface is seen to be covered with fine cusps, which confirms that the micromechanism of fracture is by microvoid

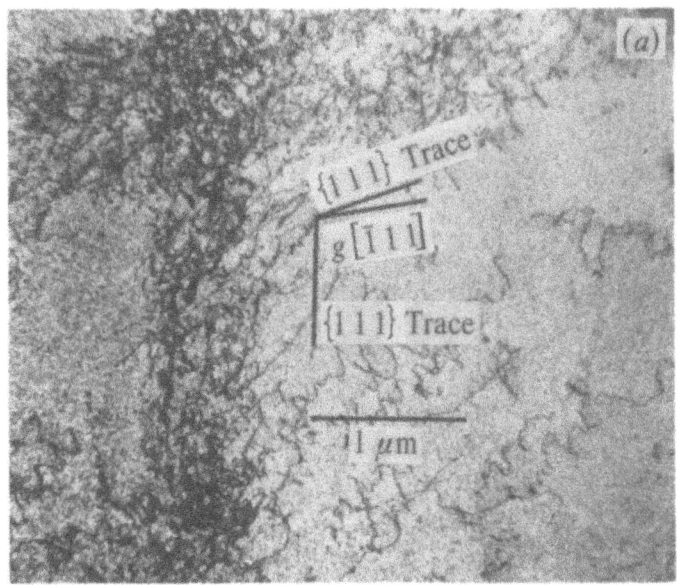

Fig. 3.16. Electron micrographs showing dislocation distributions in plastically strained Al-0.6wt% Mg-1wt% Si alloys in the peak-aged condition: (*a*) ternary alloy showing intense bands of dislocations; (*b*) alloy containing 0.5wt% Mn, dispersoid of α-$Al_{12}Mn_3Si$-phase homogenizes the dislocation distribution. (After Dowling & Martin, 1973.)

coalescence associated with particles in the grain boundaries. Such voids are nucleated at small applied strain since the inhomogeneous deformation mode coupled with a coarse grain size results in high local strains where the intense slip bands meet the grain boundaries.

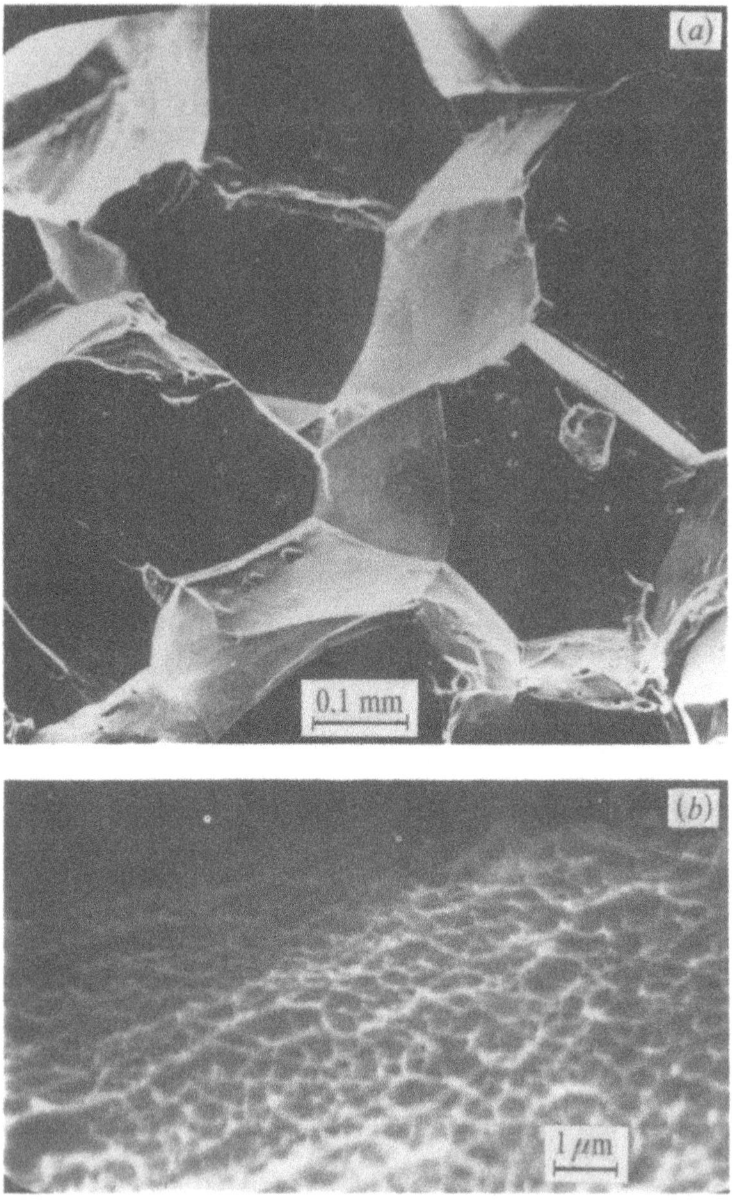

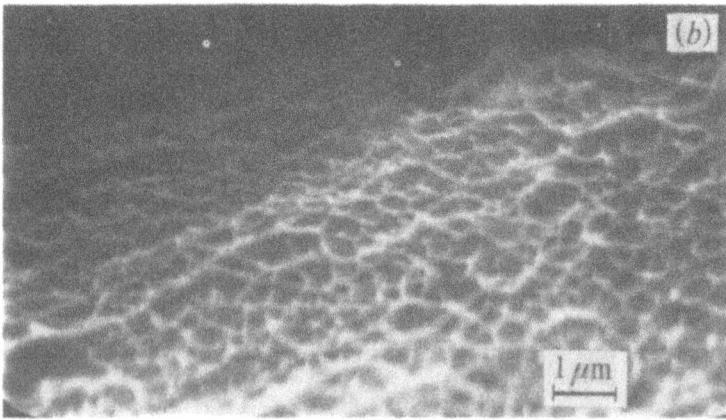

Fig. 3.17. Scanning electron micrographs of fracture surface of Al–0.6wt% Mg–1wt% Si alloy aged to peak hardness: (*a*) showing intergranular nature of fracture path; and (*b*) at higher magnification showing presence of ductile cusps. (After Dowling & Martin, 1973.)

The presence of the Mn-intermetallic dispersoid in the other specimen shown in fig. 3.16 caused an increase in the elongation to fracture to about 13%. The number of dislocations per slip band is reduced (both because of the homogenized slip and also because the dispersoid maintains a finer grain size in the material), so that failure will not be nucleated until considerably higher macroscopic strains have been applied. It should be emphasized that the relative importance, in this type of fracture, of the grain size and of the slip distribution is not unambiguously established. It is, however, likely that grain-size effects may dominate the process.

Propagation of intergranular fibrous fracture

This process is associated with the presence of precipitate-free zones (PFZs) adjacent to the grain boundaries (see chapter 1). The macroscopic strain at which this intercrystalline fracture process occurs will depend upon the difference in strength between the PFZ and the grain interiors. Thus, if overageing lowers the strength of the matrix sufficiently, transcrystalline fracture and increased ductility will be observed. We will consider the situation where the grain interiors are so much stronger than the PFZs that the fracture is confined to the latter.

Kawabata & Izumi (1976) have recently developed a theory to account for the observed fracture strain in an Al-Zn-Mg alloy aged to contain PFZs of various widths. Their starting-point is an empirical expression due to Gurland & Plateau (1963) for the fracture strain ϵ_f associated with void growth and coalescence in a solid containing a uniform volume fraction f of inclusions,

$$\epsilon_f = k\,(1 - f/f), \tag{3.18}$$

where k is a constant (this may be compared with (3.9)).

It is assumed that the fracture strain in the PFZ corresponds to the macroscopic fracture strain calculated by (3.18). The volume fraction of the grain-boundary precipitates in the PFZ (f_{gbp}) is given by

$$f_{gbp} = k'\,d^3\,N_S/w. \tag{3.19}$$

$k'd^3$ is the volume of one precipitate of diameter d, where k' is a shape factor. N_S is the number of grain-boundary precipitates per unit area, and w is the width of the PFZ. Substituting (3.19) into (3.18) we obtain

$$\epsilon_f = k''\,(w/d^3 N_S) - 1,$$

where k'' is another constant. Since $w/d^3 N_S \gg 1$, this may be simplified to

$$\epsilon_f = k''\,w/d^3 N_S. \tag{3.20}$$

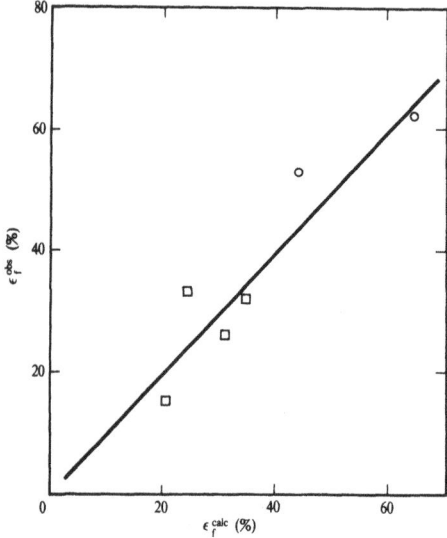

Fig. 3.18. Calculated fracture strain, ϵ_f^{calc} (with k'' assumed to be 0.134), versus observed fracture strain, ϵ_f^{obs}. (After Kawabata & Izumi, 1976.)

Kawabata & Izumi compared calculated values of ϵ_f based on (3.20) with experimentally determined fracture strains ($\ln A_0/A_f$) for a series of alloys that had been heat-treated to produce grain-boundary PFZ widths varying from 0.042 μm to 0.126 μm, and particle diameters of the order 0.1 μm within the grain boundaries. Their results are shown in fig. 3.18, and it can be seen that, although the data scatter a little, there appears to be reasonable support for (3.20) in spite of the simplifications involved and the neglect of nucleation strain for the formation of the voids at the grain-boundary particles.

3.6   Fracture toughness of particle-hardened alloys

We have so far considered fracture processes in cylindrical specimens subjected to longitudinal tensile stress, therefore involving both the nucleation and propagation of the fracture. The integrity of any engineering structure depends, however, upon the fracture toughness of the material from which it is made. Assessing this property involves establishing the conditions under which a sharp crack begins to propagate through the material. We will proceed to consider the extent to which it is possible to relate macroscopic measurements of fracture toughness to the local micromechanisms of fracture. First, the various fracture toughness parameters will be defined.

3.6.1   Fracture toughness parameters

Linear elastic crack tip field

In an infinite body containing a central, through-thickness crack of length $2a$, the Griffith equation states that elastic fracture will occur at a critical value of the potential energy release rate per unit thickness,

$$G_c = 2\sigma, \tag{3.21}$$

where 2σ is the elastic 'work to fracture', being the surface energy per unit area.

An alternative form of (3.21) makes use of the relationship between stress intensity factor, K, and energy release rate G. For tensile loading across the crack faces (mode I) this relationship is

$$G = \alpha K^2 / E, \tag{3.22}$$

where $\alpha = 1$ in plane stress and $(1 - \nu^2)$ in plane strain. ν is Poisson's ratio and E is Young's modulus. The critical value of stress intensity factor corresponding to (3.21) is

$$K_c = p_f (\pi a)^{\frac{1}{2}}, \tag{3.23}$$

where p_f is the fracture stress of the body. Substituting (3.23) into (3.21) yields

$$p_f = (EG_c / \pi a)^{\frac{1}{2}}. \tag{3.24}$$

Nonlinear behaviour at the crack tip

Most metals do not fail in a completely elastic manner, but only after some local yielding has occurred at the crack tip. Provided that the non-elastic region is small compared with the dimensions of the specimen and of the crack, it is possible to calculate the potential energy release rate using elastic theory. Much of the energy absorption in the form of strain-energy in the dislocations in the plastic zone, is concerned with this local plasticity. Equation (3.21) thus becomes

$$G_c = 2\sigma + \sigma_p, \tag{3.25}$$

where σ_p is a plastic work term.

The size and shape of the plastic zone at the tip of a propagating crack can be calculated, and fig. 3.19 illustrates schematically such a zone under conditions of (*a*) plane stress (e.g. a crack in a thin sheet) and (*b*) plane strain (e.g. as at a crack tip in the centre of a thick sheet or plate). In fig. 3.19*b*, the size of the zone (r_Y), which extends as two lobes inclined at about 70° to the line of crack extension, depends on the yield stress of the

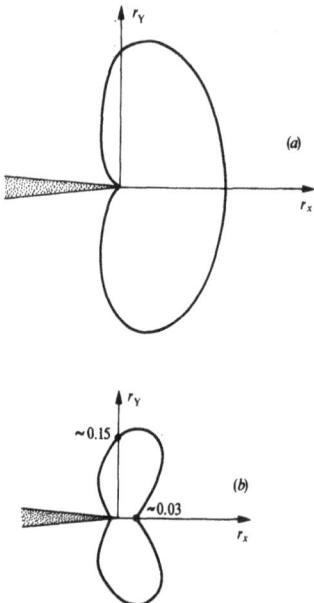

Fig. 3.19. The shape of the plastic zone at the tip of a propagating crack under conditions of: (*a*) plane stress; and (*b*) plane strain.

metal (p_Y) and the stress intensity factor K at the tip of the crack,

$$r_Y = A\,(K/p_Y)^2, \tag{3.26}$$

the approximate values of the constant A being given in fig. 3.19b. The dimensions of standard test-pieces are specified in such a way that the size of the plastic zone is very much smaller than the distance the crack has to propagate (the 'ligament width') in the uncracked specimen, so that linear elastic analyses may be acceptably employed. A second requirement of the test-piece is that its thickness is sufficient to ensure plastic constraint at the crack tip, to provide conditions of plane strain. Under these conditions the critical stress intensity for crack propagation under mode I stressing is termed K_{Ic}. The size of the plastic zone is seen to be dependent upon the yield stress of the material (3.26), and in order to ensure that the toughness does correspond to 'plane strain' the standards specify that the specimen thickness B is greater than $2.5\,(K_{\mathrm{Ic}}/p_Y)^2$.

A further feature of a yielded crack is that there is a discrete opening at the crack tip which does not occur for an elastic crack. This opening is known as the *crack opening displacement* (COD), and an alternative fracture criterion can be postulated if it is supposed that fracture occurs at a critical COD, δ_c. If G_c is calculated from the incremental work done per unit thickness for a crack advance, δa, achieved by a force per unit thickness of $p_Y \delta a$, moving through a displacement δ_c, we obtain

$$G_c = p_Y \, \delta_c. \tag{3.27}$$

Thus, from (3.22), we have

$$\delta_c = \alpha K_c^2 / p_Y E. \tag{3.28}$$

In attempting to correlate micromechanisms of fracture with toughness parameters in materials exhibiting small-scale yielding at the crack tip, the critical size of the plastic zone may be that necessary to build up sufficient tensile stress ahead of the crack tip to propagate a brittle crack nucleus. δ_c may be associated with the coalescence of voids formed around second-phase particles to give ductile rupture. We will consider these mechanisms in turn.

3.6.2   Cracking processes

Mild steel

As discussed in §3.4.1, growth-controlled cleavage fracture in mild steel is believed to be nucleated by the cracking of grain-boundary carbide particles. It has proved possible to calculate the plane strain fracture toughness, K_{Ic}, for this material for a range of low temperatures over which values of the cleavage fracture stress p_f had been determined.

At any temperature, the plastic zone ahead of the pre-crack in the toughness test-piece has to develop to a size sufficient to raise the maximum tensile stress in the zone to the value required to propagate the nucleus. At low temperatures the yield stress p_Y is high, so the plastic zone can be small, and the toughness is low. With increasing temperature and decreasing yield stress the size of the plastic zone and thus the toughness increase.

The distribution of tensile stress ahead of the crack tip in, for example, a toughness test-piece has been calculated, and it is of the form indicated in fig. 3.20a. The scale of the ordinate has been normalized with respect to the yield stress, and the abscissa is the ratio of distance ahead of the crack tip, x, to plastic zone size, r_Y, where $r_Y = A(K^2/p_Y^2)$ as in (3.26). This means that as the plastic zone size increases, so the absolute value of distance x increases, and a given stress level is produced at a *greater* distance from the tip of the crack.

Ritchie, Knott & Rice (1973) assumed that the critical plastic zone size is that required to give a stress p_f at a characteristic distance x_c ahead of the crack tip. Since it is necessary to find a nucleus to propagate, suitable distances were thought to consist of one or two grain diameters, because

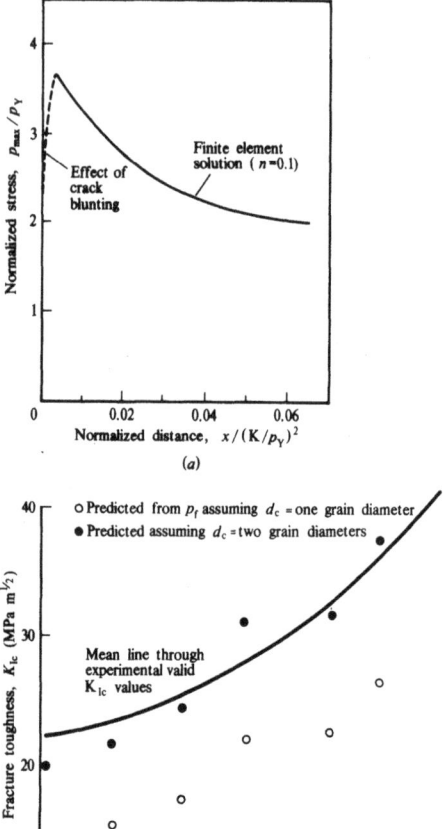

Fig. 3.20. (*a*) The stress distribution ahead of a blunting crack (after Knott, 1977). (*b*) Comparison between experimental values of K_{Ic} and those predicted from p_f as a function of temperature for a high nitrogen steel (after Ritchie, Knott & Rice, 1973).

a nucleus consists of grain-boundary carbide.

The continuous line in fig. 3.20*b* illustrates the fracture toughness as a function of the temperature of a high-nitrogen steel of grain size 60 μm. The value of p_f was determined as 850 MPa, this value being assumed to be independent of temperature over the range investigated. Measurements of the yield stress were then made over the same temperature range, and from fig. 3.20*a* and (3.26) it is possible to calculate values of K_{Ic} assuming various fixed values of critical distance (x_c). The results are shown in fig. 3.20*b*, and they indicate the correct general trend, although the extremely good agreement with a characteristic distance of two grain diameters is unlikely to be other than fortuitous.

Spheroidized carbon steels

Rawal & Gurland (1977) have recently explored the effect of cementite particles on the fracture toughness of spheroidized carbon steels. The carbon content of their steels ranged from 0.13 to 1.46wt%, and they determined yield strengths and K_{Ic} values of these materials at −198, 170, −150 and −110 °C. At the lower two temperatures the specimens were observed to fracture in a brittle manner by cleavage, increasing amounts of fibrous fracture being observed at the higher temperatures.

Cleavage crack initiation was found to be associated with the carbide particles. This was to be expected since the maximum stress intensity ahead of the crack tip was found almost always to exceed the expected stress for particle fracture or decohesion. Rawal & Gurland went on to apply to their data the analysis of Ritchie, Knott & Rice described in the previous section. The characteristic distance selected was 1.3 times the ferrite grain size, and fig. 3.21 illustrates the degree of correlation between the calculated and measured K_{Ic} values they obtained.

3.6.3   Ductile rupture processes

Transgranular rupture

Hahn & Rosenfield (1975) have developed a model for ductile failure, having recognized that voids can form at very low strains in association with large (i.e. up to 10 μm in diameter) inclusions in metals. Their model

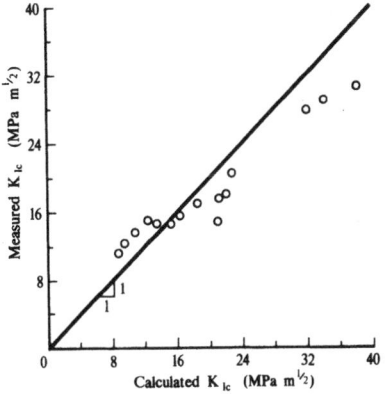

Fig. 3.21. Comparison of measured K_{Ic} and calculated K_{Ic} for spheroidized carbon steels. (After Rawal & Gurland, 1977.)

is based on the concept that the critical step in the crack extension process occurs when the extent of the heavily deformed region at the crack tip is comparable with the width of the unbroken ligament separating the cracked particles. Rice & Johnson (1970) have calculated that the region of large plastic deformation at the crack tip extends a distance comparable to δ, the crack opening displacement. Thus if λ_c is the spacing of the cracked particles, the Hahn & Rosenfield criterion may be written as

$$\lambda_c = \delta_c. \qquad (3.29)$$

If the mean volume spacing of the particles is expressed in terms of their volume fraction f and diameter d i.e. $\lambda_c = d(\pi f/6)^{\frac{1}{3}}$, substitution in (3.28) yields

$$K_{Ic} = [2\,p_Y\,E\,(\pi/6)^{\frac{1}{3}}\,d]^{\frac{1}{2}}\,f^{-\frac{1}{6}}. \qquad (3.30)$$

Fig. 3.22 indicates the application of (3.30) to data for a number of materials containing coarse inclusions, notably a range of commercial aluminium alloys and steels. There appears to be good experimental agreement with the predicted linear relationship between K_{Ic} and $f^{-\frac{1}{6}}$, but certain other implications of this equation throw some doubt upon its validity, and we will discuss these points below. Nevertheless the same approach has been used by Rawal & Gurland (1977) to account for the effect of cementite particles upon the fracture toughness of spheroidized carbon steels referred to in §3.6.2. When these steels were fractured at temperatures leading to fibrous failure (rather than under cleavage conditions as referred to earlier), the final void size was about equal to the planar nearest-neighbour particle spacing. The microstructure of the steels satisfied the geometrical condition for void coalescence to a planar void

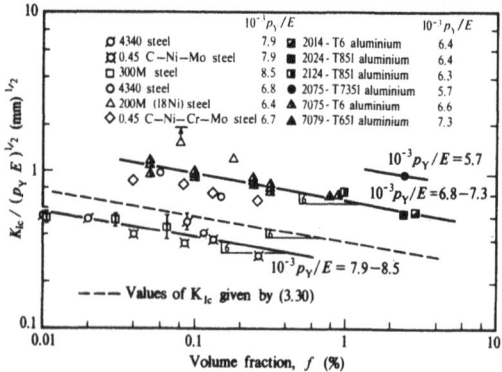

Fig. 3.22. Influence of the volume fraction, f, of large ($\geqslant 1\ \mu$m) size second-phase particles on the plane strain fracture toughness of commercial alloys. (After Hahn & Rosenfield, 1975.)

sheet, so λ_c is now more appropriately described as the mean planar nearest-neighbour particle spacing, i.e. $\lambda_c = 0.36\, d\, f^{-\frac{1}{2}}$. Substitution into (3.28) yields a relation of the form

$$K_{Ic} \propto p_Y^{\frac{1}{2}} f^{-\frac{1}{4}}. \tag{3.31}$$

Rawal & Gurland's results are illustrated in fig. 3.23 for a number of spheroidized steels plotted in accordance with (3.31), and again good agreement is observed.

The first shortcoming of (3.30) and (3.31) is that they predict increases in fracture toughness with increasing yield stress, p_Y, at constant volume fraction and distribution of the inclusions. In the case of age-hardening aluminium alloys which exhibit transgranular fracture, the toughness decreases as the yield strength is increased.

The second criticism of (3.30) is that toughness does not increase with increasing size, d, of void-nucleating particles. The opposite effect has been found by Cox & Low (1974) in comparing steels of the same yield strength which contained identical inclusion spacings differing only in size. They found that the smaller inclusion sizes gave rise to substantial increases in toughness. Furthermore, for a constant inclusion dispersion, specimen orientations with the largest particle size normal to the stress axis (short transverse) typically have substantially lower toughness values, which is contrary to the prediction of (3.30). Specimens cut in the longitudinal orientation, where inclusions are elongated normal to the line of crack advance, show high values of δ_c and thus K_{Ic}. The increase

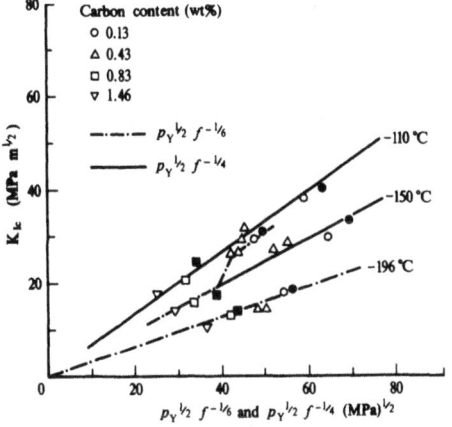

Fig. 3.23. Influence of the volume fraction of carbide particles and yield strength on the plane strain fracture toughness of spheroidized carbon steels. (After Rawal & Gurland, 1977.)

occurs because it is necessary to spread high strains longitudinally as far as the ends of the inclusions.

Effect of work-hardening and slip distribution. With increasing local crack-tip plastic strains, some materials may lose their capacity for continued work-hardening and flow begins to localize, eventually producing shear fracture along localized flow bands. The orientation of these bands of shear fracture will depend upon the crack-tip sharpness at the point when the plastic flow becomes localized, as observed by Clayton & Knott (1976), see fig. 3.24.

In a material of low work-hardening capacity (e.g. a cold-worked steel), shear fracture will propagate with negligible crack-blunting. The slip-line field ahead of a sharp crack is as indicated in fig. 3.24a. If voids have nucleated at low strain around inclusions, then shear bands will become localized in the directions of the nearest holes ahead of the crack tip (fig. 3.24b), and it is possible for decohesion to occur by shear along these bands before the crack tip has blunted appreciably. The final fracture will be crack-like, and zig-zags from inclusion to inclusion (fig. 3.24c). Such

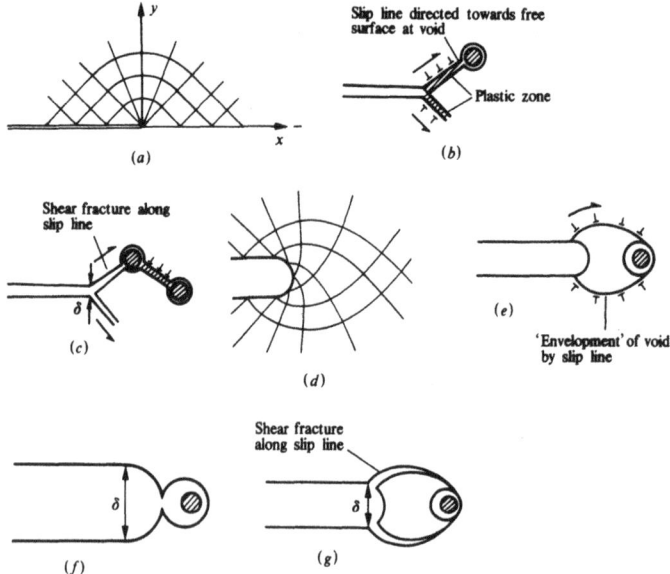

Fig. 3.24. Shows: (a) local plane strain slip-line field around a crack tip; (b) direction of shear band towards void in non-hardening material; (c) decohesion along shear band; (d) modification of slip-line field due to crack-tip blunting; (e) envelopment of void by logarithmic spiral slip line; (f) void coalescence by internal necking in hardening matrix; (g) development of (e) to give shear cohesion. (After Clayton & Knott, 1976.)

fracture paths have been observed in wrought steels.

If the material does not exhaust its work-hardening capacity, and flow is not localized until after the crack tip has blunted significantly, the slip-line field will be as in fig. 3.24d, which has the form of logarithmic spirals. Once a void at an inclusion ahead of the crack tip is 'enveloped' by a slip line (fig. 3.24e), normal void coalescence by internal necking will proceed if flow is not localized by exhaustion of work-hardening (fig. 3.24f). If flow does become localized, shear fracture will take place along the spiral slip-line path (fig. 3.24g).

The effect of work-hardening capacity and its effect upon flow localization is the basis of a simple model due to Hahn & Rosenfield (1968) which attempts to relate the fracture toughness parameter to the unidirectional tensile properties and other microstructural parameters. A material having a high capacity for strain hardening (high n value, where the form of the tensile stress-strain curve is given by $p = p_0 \epsilon_p{}^n$) will effectively have a *lower strain concentration* at the crack tip for a given stress level. Work-hardening essentially causes lateral spreading of the zone of plasticity into a more diffuse region, thereby dispersing the crack-tip displacements. Assuming ductile fracture will occur when a given critical strain is reached, higher stress levels can be supported before failure occurs, i.e. the material has a higher toughness. Hahn & Rosenfield write the critical crack opening displacement, δ_c, in terms of a critical fracture strain ϵ_c^*, and the plastic zone width, l^*, at the onset of fracture as

$$\delta_c = 2 \epsilon_c^* l^*. \tag{3.33}$$

Experimentally, they observed that the width of the plastic zone is proportional to n^2, and development of the model then gives

$$K_{Ic} \propto n \, (p_Y \epsilon_c^*)^{\frac{1}{2}}. \tag{3.33}$$

ϵ_c^* will be dependent upon the volume fraction of void-nucleating particles, and Garrett & Knott (1976) have found (3.33) to give quite good agreement with measured values of K_{Ic} in a range of commercial and pure aluminium alloys (fig. 3.25).

Intergranular rupture

During an isothermal ageing sequence many precipitation-hardening alloys exhibit a condition in which the interior of the grains is highly hardened while the grain boundary and its environment (the PFZ) is very soft. This tendency, and thus the inhomogeneity of distribution of strain between grain interior and PFZ will be maximized close to the maximum of the yield stress of the bulk alloy. As discussed earlier (§3.5.2) plastic strain can be highly localized inside the soft PFZs, and they become preferred

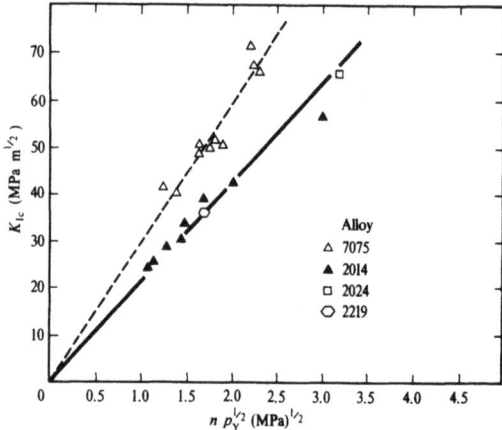

Fig. 3.25. A comparison of the predicted (using (3.33)) and observed relation between fracture toughness, K_{1c}, and the tensile parameter $n/p_Y^{\frac{1}{2}}$. (After Garrett & Knott, 1976.)

sites for initiation and propagation of cracks. Hornbogen (1975) considered how the Hahn & Rosenfield approach of (3.33) might be modified to describe the fracture toughness of materials undergoing crack propagation within the PFZ.

This approach involves defining a critical fracture strain ϵ_{ci}^* for localized intergranular failure. Fig. 3.26 illustrates schematically how the localized plastic deformation in the vicinity of grain boundaries may be defined. For material of grain size D subjected to a shear displacement x, the homogeneous shear strain is x/D. If the strain is accommodated by grain-boundary sliding, then the local shear strain is x/b, where b is the width of the grain boundary. If there is a PFZ of width d, however, then the local shear strain due to sliding in the PFZ may be written as x/d. It can thus be seen that the critical *macroscopic* strain will depend upon the parameters d and D, and one may write

$$\epsilon_c^* = \epsilon_{ci}^* \, d/D. \tag{3.34}$$

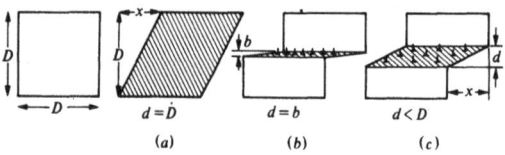

Fig. 3.26. Schematic representation of localized plastic deformation associated with grain boundaries. (After Hornbogen, 1975.)

Substitution into (3.33) gives

$$K_{Ic} \propto n_i \, (p_{Yi} \epsilon_{ci}^* \, d/D)^{\frac{1}{2}},$$

i.e. $K_{Ic} = kD^{-\frac{1}{2}},$ (3.35)

where n_i is the work-hardening exponent for the grain-boundary PFZ of constant thickness, and k is a constant. Equation (3.35) indicates that a $D^{-\frac{1}{2}}$ grain-size dependence of fracture toughness can be expected if deformation takes place in a PFZ of constant thickness. Fig. 3.27 shows the grain-size dependence of the fracture toughness K_c (i.e. not under plane strain conditions) measured with a precipitation-hardening 7075 aluminium alloy (Al–Zn–Mg–Cu) which showed denuded zones at the grain boundaries. The alloy also exhibited preferred deformation in the PFZs and fractured intergranularly, and although there is a considerable scatter in the data, the toughness variation is broadly consistent with the Hornbogen model.

It appears probable, therefore, that a more exact quantitative correlation of microstructural features and macroscopic crack-propagation behaviour will be obtained for particle-hardened alloys if the inhomogeneity of strain is considered. The intergranular failure strain, ϵ_{ci}^*, is likely to be dependent upon the volume fraction f_i of particles in the grain boundaries. In most precipitation-hardened alloys, incoherent particles form by heterogeneous nucleation (see chapter 1), and by analogy with (3.9) one may write

$$\epsilon_{ci}^* = \frac{1}{C} \ln \left[\sqrt{(\pi/6f_i)} - \sqrt{\tfrac{2}{3}} + e_n \right].$$ (3.36)

Many microscopic investigations have shown that the local volume fraction, f_i, can be very high, which then leads to small values of ϵ_{ci}^*, even in a very soft and ductile matrix.

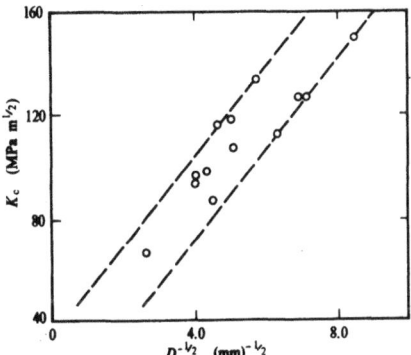

Fig. 3.27. Grain-size dependence of the fracture toughness, K_c, of the precipitation-hardened 7075 aluminium alloys. (After Hornbogen, 1975.)

3.7   Fatigue fracture of particle-hardened alloys

It is well established that the age-hardening process in many alloys, which has a large effect upon the tensile strength, may have relatively little effect upon the fatigue strength. This results in a low value of the endurance ratio, i.e., the ratio of the fatigue stress giving failure in 10^7 or 10^8 cycles to the ultimate tensile strength. Thus, fig. 3.28 illustrates the variation of the 10^8 cycles endurance limit with UTS for different aluminium alloys. This shows that for annealed alloys a ratio of about 0.5 is obtained, which is similar to that obtained in pure metals. Cold-working gives rise to a ratio of 0.33 and age-hardening to a ratio of only 0.25. The value of the endurance ratio for a given alloy changes during heat-treatment, and in the case of a Al–Zn–Mg–Cu alloy, Forsyth (1963) reports values of 0.4, 0.25 and 0.45 for the solution-treated, aged to peak-hardness and overaged conditions, respectively. A very similar series of ratios is reported by Bonfield (1972) for an age-hardening Cu–1.9wt% Be alloy in similar metallurgical conditions. High-strength steels which have been hardened by fine dispersions of carbide or by the precipitation of intermetallic phases can also show low endurance ratios. Some clarification of the phenomenon may be obtained by conducting static mechanical tests upon fatigued specimens, as discussed below.

3.7.1   Cyclic softening effects

The response of materials to strain cycling may be assessed by measurement of the stress required to produce a given plastic strain at different stages of the fatigue life. Annealed single-phase materials usually harden up to a plateau value of stress which is then maintained for the rest of the life, whilst work-hardened materials soften due to instability of the dislocation structure under fatigue stressing. Laird (1976) observed that cyclic

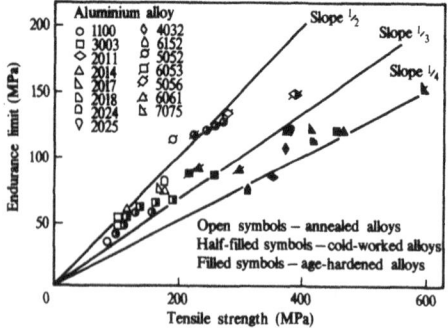

Fig. 3.28. Variation of 10^8 cycles endurance limit with tensile strength for different aluminium alloys. (After McEvily et al., 1963.)

softening is qualitatively understood in terms of dislocation rearrangements in those metals where strength is associated with dense populations of dislocations. It is not understood how these rearrangements take place, nor can one predict accurately what the flow stress should be as a result of cycling. The situation in alloys hardened by precipitates is even more difficult, and there are several theories about the nature of the softening mechanisms involved.

Smith (1975) has recently reviewed the interrelation of the microstructure and the fatigue properties of precipitation-hardening aluminium alloys, and much of what follows is derived from his article. Smith points out that many alloys containing fine precipitates, such as quench-aged Fe-C, nickel alloys, and high-strength steels as well as aluminium alloys, may show cyclic softening and the effect is associated with instability of the precipitates and/or the dislocation structures. Fig. 3.29 refers to Al-Cu, and illustrates the cyclic response curves for the material when aged to contain the coherent θ''-phase. Initial cyclic hardening up to a maximum or saturation value is followed by marked softening to failure.

One explanation of the phenomenon is that repeated shearing of the coherent particles reduces their size until they become thermodynamically unstable, so that they undergo reversion. Those regions where the precipitates have returned into solution will thus be soft zones and would account for the macroscopic loss of strength. An alternative hypothesis is that softening occurs simply by the progressive cutting or 'scrambling' of the precipitate particles, and no actual re-solution takes place. These questions have not been resolved totally satisfactorily, but the changes induced by fatigue need not always be of the same type in all materials, so re-solution, overageing, disordering of ordered precipitates and possibly dislocation-enhanced solubility may all occur.

Fine & Santner (1975) have explored the effect of various types of dispersion in an aluminium alloy upon cyclic softening. Three different

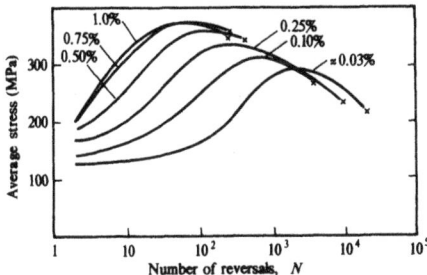

Fig. 3.29. Cyclic response curves at different ϵ_p values for Al-Cu alloys aged to contain the θ''-phase. (After Calabrese & Laird, 1974.)

Al-Cu alloys were chosen: Al-3.6wt% Cu; Al-6.3wt% Cu; and a commercial 2024 alloy. Each alloy was tested in the naturally aged condition, which allowed a comparison of a duplex phase structure with a single precipitate structure. The Al-3.6wt% Cu had only GPI present, the Al-6.3wt% Cu had both GPI and undissolved 5 to 10 μm diameter θ particles, and the commercial 2024 had GP zones, dispersoids and inclusions present.

Fig. 3.30 illustrates the changes observed in the peak (engineering) stress sustained by the three alloys as a function of the number of strain reversals during strain-controlled cycling at a number of strain amplitudes. The precipitous load drop does not arise from cyclic softening, but is due to the formation of cracks in the material. As the crack grows, the maximum tensile load continuously decreases, going down almost to zero before final failure.

In fig. 3.30a the alloy with only GPI zones present shows cyclic hardening followed by cyclic softening at all strain levels, which is behaviour consistent with the results of Calabrese & Laird (fig. 3.29). The

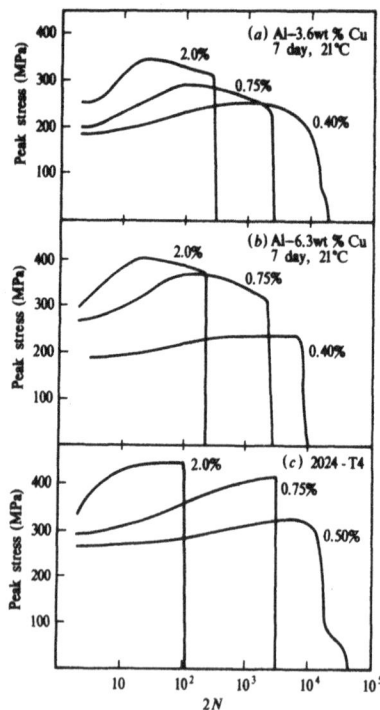

Fig. 3.30. Peak (engineering) tensile stress versus the number of reversals, N, during strain-controlled cycling at a number of strain amplitudes for three aluminium alloys. (After Fine & Santner, 1975.)

2024 alloy, which has dispersoids and inclusions present, shows (fig. 3.30*c*) cyclic hardening followed by saturation at all strain levels tested. The Al-6.3wt% Cu alloy (fig. 3.30*b*) shows cyclic hardening followed by softening at large and intermediate strains, but no softening was observed at low strains.

These data demonstrate that the micromechanism responsible for cyclic softening is dependent both on the microstructure and upon the slip distribution in the material. The changes in fig. 3.30*a* arise from a scrambling or disordering of the structure in fatigue. When a GP zone is cut by a dislocation there is a decrease in the number of Cu–Cu bonds and an increase in the number of Al–Cu bonds, i.e. the structure becomes more random. The smaller cyclic softening in fig. 3.30*b* is attributed to the large θ particles causing the plastic deformation to be less localized: the structure begins to disorder at higher strain amplitudes. The distribution of dispersoids and inclusions in 2024 alloy causes sufficient slip homogenization to prevent cyclic softening at all the strain ranges tested.

3.7.2   Structural changes during fatigue

Certain inhomogeneities of microstructure will exist in age-hardened alloys as a result of their heat-treatment. For example, grain-boundary PFZs may be present and these, as in the case of tensile deformation, will give rise to localized deformation and intergranular fracture in fatigue. Transcrystalline ageing inhomogeneities will also be present, and these have been the subject of much speculation in relation to their effects on fatigue. These can arise from quench bands (i.e. preferential nucleation upon dislocations generated by differential contraction strains from the quench), and in Al–Cu alloys, coarse local regions of θ' interspersed with fine θ'' are found. These features can act as soft zones, but they do not completely explain the low fatigue strengths of age-hardened alloys, since further structural changes may be brought about by the fatigue strain itself.

Bands containing dense arrays of dislocations have been found after the fatigue of a pure Al–Cu containing θ'-phase (Calabrese & Laird, 1974). There were several bands per grain visible throughout the cross-section after etching, but electron microscopy showed no overageing or re-solution and it is suggested that they are fatigue-induced soft zones caused by a repeated scrambling of the atoms in the precipitates by shear. Quench-aged Fe–C containing a fine dispersion of metastable coherent carbides shows cyclic softening and well-defined precipitate-free channels after fatigue (McGrath & Bratina, 1967), whilst an Fe–Cu alloy containing fine noncoherent particles of Cu does not show cyclic softening, or any evidence for the formation of precipitate-free regions (McGrath & Bratina, 1970). Assuming that in both cases the particles are sheared, the difference could be due to local diffusion rates in regions of concentrated slip controlling whether re-solution (or even overageing) can occur.

3.7.3   Fatigue crack nucleation in precipitation-hardened alloys

Precipitation-strengthened alloys do not appear to exhibit essentially different mechanisms of formation of fatigue crack nuclei from pure metals, when transgranular crack nucleation is considered. Naturally, crack initiation in commercial alloys can be from coarse inclusions, which are often in the form of intermetallics produced by impurity elements (e.g. due to iron in Al alloys). The local stress concentration at the inclusion induces concentrated slip in the adjacent matrix, leading to the formation of fatigue crack nuclei.

The influence of the homogeneity of slip upon fatigue properties can be seen in the case of aluminium alloys by comparing the underaged condition of a commercial alloy with its pure version. Insofar as fatigue cracks are associated with the appearance upon the surface of 'persistent slip bands', high slip steps in a material exhibiting inhomogeneous slip are likely to favour crack nucleation. Fig. 3.31 illustrates the S–N curves of two aluminium alloys. One is a commercial 7075 alloy and the other is a pure version of the same alloy, termed X-7075. Both have been aged to the same hardness, and contain a dispersion of coherent particles.

The commercial 7075 alloy also contains incoherent dispersoids due to the presence of a transition element (as in the alloy of fig. 1.1), which, as previously discussed have the effect of homogenizing the slip distribution. The pure alloy, X-7075, on the other hand, displays inhomogeneous slip because of the mechanism of shearing the coherent hardening phase, and it

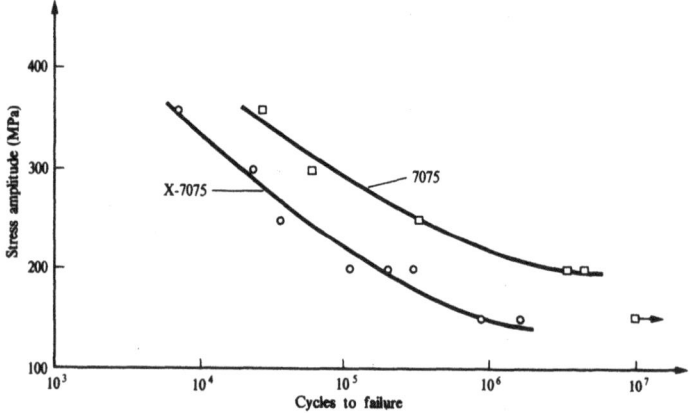

Fig. 3.31. S–N curves for commercial 7075 alloy and the pure equivalent (X-7075), each aged 24 hours at 100 °C. (Courtesy of G. Lütjering.)

is seen that this alloy has a considerably worse fatigue life than the commercial material. Even in fatigue deformation, the dislocation distribution of the 7075 alloy remains homogeneous as a result of the presence of the hard and undeformable inclusions.

The greater part of the added elements manganese, silicon, chromium and iron that are to be found in commercial aluminium alloys are present in the form of inclusions. In the present discussion it is clear that it is important to make a distinction between the relatively coarse inclusions ($> 5~\mu$m) that are to be found in almost all alloys and which are always deleterious in fatigue (and to fracture toughness as we have already discussed), and the small dispersoids (0.1 to 1.2 μm) in commercial alloys which are of importance in influencing the slip distribution (as well as in inhibiting grain growth).

3.7.4   Fatigue crack propagation in precipitation-hardened alloys

The application of linear elastic fracture mechanics and related small-scale crack-tip plasticity have provided the basis for describing the phenomenon of fatigue-crack propagation. Many investigators have confirmed that the crack growth rate per cycle (da/dN) is primarily controlled by the alternating stress intensity (ΔK) through an expression due to Paris & Erdogan (1963) of the form

$$da/dN = C\Delta K^m, \tag{3.37}$$

where C and m are scaling constants, and ΔK is given by the difference between the maximum and minimum stress intensities for each cycle, i.e. $\Delta K = K_{max} - K_{min}$. Equation (3.37) provides a reasonable description of growth rates in the range of approximately 10^{-5} to 10^{-3} mm per cycle, but at higher growth rates, when K_{max} approaches K_{Ic} (the fracture toughness), (3.37) underestimates the crack propagation rate. At lower growth rates (3.37) is found to be conservative, because ΔK approaches a threshold stress intensity range ΔK_0, below which crack propagation cannot be detected.

Lindley, Richards & Ritchie (1975) have surveyed the types of fatigue fracture surfaces of metals having a wide variety of microstructures and tested over a wide range of ΔK and K_{max}. They conclude that there are four general methods of growth, namely striation formation, cleavage, void coalescence and intergranular separation. The plot of fatigue growth rate against ΔK is, for the reasons discussed above, sigmoidal as shown in fig. 3.32, and the three regimes of the curve can be characterized in terms of different primary fracture mechanisms. In the mid-ΔK range (regime B in the figure) where failure generally occurs (for example in steels) by a transgranular ductile striation mechanism, there is often little influence of

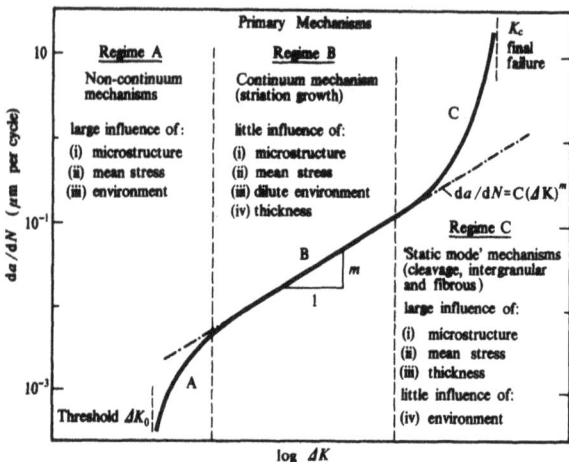

Fig. 3.32. Summary diagram showing the primary fracture mechanisms associated with the sigmoidal variation of fatigue-crack propagation rate with alternating stress intensity. (After Lindley, Richards & Ritchie, 1975.)

microstructure and mean stress on crack growth. Specimen thickness, and hence degree of plastic constraint and a dilute environment also appear to have little effect upon the crack propagation rate. At higher growth rates (regime C in the figure) when K_{max} approaches K_{Ic}, growth rates become extremely sensitive to both microstructure and mean stress, due to a departure from striation formation to include 'static' fracture modes, such as cleavage and intergranular or fibrous fracture. In regime A (fig. 3.32), as ΔK_0 is approached, there is similarly a strong influence of microstructure and mean stress on growth rates, together with an increased sensitivity to environmental effects.

When micromechanisms of fatigue-crack growth are considered, a quantitative description can be given of the process, based on the plastic shear that occurs at the crack tip under the condition that deformation is limited to crystallographic slip in a single crystal. In this case (Neumann, 1974), a full correlation between fatigue-crack growth and the microscopic slip steps is achieved. If, however, real metallic materials which always contain grain boundaries and particles are considered, the situation becomes very complex. For this reason, the design of fatigue-resistant alloys is still done on an empirical basis.

Hornbogen & Zum Gahr (1976) have recently considered some of the micromechanical parameters that may be effective at the crack tip, and suggest the following factors as likely to have an effect on the propagation of fatigue cracks:

(i) the heterogeneity of slip steps within the matrix;

(ii) localized slip in grain boundaries or PFZs;
(iii) decohesion of cleavage planes in matrix or particles;
(iv) decohesion of slip planes in the matrix;
(v) decohesion of incoherent particle/matrix interfaces;
(vi) decohesion of embrittled grain boundaries.

They point out that the ductile mechanisms are expected to prevail at low ΔK values, while at higher stress intensities, R-values and specimen thicknesses, an increasing amount of quasi-static cleavage may contribute to crack propagation.

Hornbogen & Zum Gahr examine the effect of microstructure on the growth of fatigue cracks in a γ-Fe–Ni–Al-alloy for different conditions of precipitation. Heat treatments were chosen in such a way that the alloy deformed by homogeneous or heterogeneous plastic deformation in spite of identical tensile properties. This was effected by having specimens in both the aged and overaged conditions, so that the difference in the mode of plastic deformation was caused by particles that are sheared or bypassed by dislocations, respectively. A third microstructure was added with heterogeneous plastic strain but gave much smaller yield and tensile stresses (effected by underageing) than the other two series. Fig. 3.33

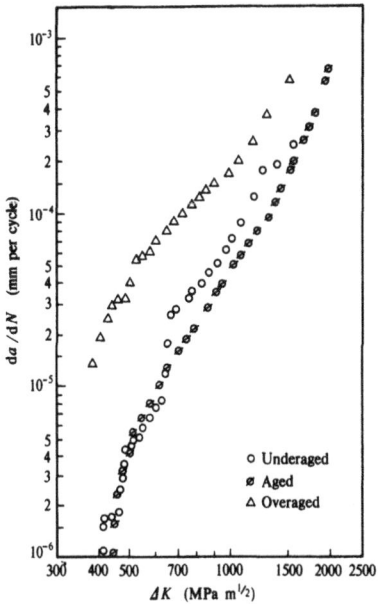

Fig. 3.33. Crack growth curves for underaged, aged and overaged Fe–Ni–Al alloy. Specimen thickness 1.5 mm, R value 0.235. Environment: air temperature, 21 to 25 °C; relative humidity: 43 to 50%. (After Hornbogen & Zum Gahr, 1976.)

illustrates some of the crack growth curves obtained for the three micro-structures as a function of stress intensity range for specimens of thickness 1.5 mm and with an R value of 0.235. All testing conditions gave essentially the same behaviour, namely that crack propagation for small ΔK was much slower for the underaged and aged specimens than for the overaged specimen. For higher ΔK, the difference decreases. The crack velocities in the underaged and aged specimens were about equal.

An increasing stress intensity leads to an increase in the size of the plastic zone (3.26) ahead of the crack. The 'static' plastic zone size, r_s, is given by

$$r_s = A \, (K_{max}/p_Y)^2,$$

and the cyclic or fatigue plastic zone size will be,

$$r_f = A' \, (\Delta K/p_Y)^2,$$

where A and A' are constants. These zone sizes may be correlated with microstructural features of the alloys. In fig. 3.33 a discontinuity occurs in the underaged and aged specimen curves where r_s is roughly equal to the grain size D (where $D = 165$ μm) and $\Delta K = 480$ MPa m$^{-\frac{1}{2}}$), and a second discontinuity in the data for the underaged specimen is found at r_f approximately equal to D, i.e. when $\Delta K = 700$ MPa m$^{-\frac{1}{2}}$. The relationship between the plastic zone sizes and the microstructure in the underaged and aged specimens is illustrated diagrammatically in figs. 3.34a, b, c and d, where the following possibilities are considered with increasing ΔK.

Stage I. $r_f < r_s < D$: dislocations are nucleated at the crack tip and travel within the interior of the grain until the external stress is less than the yield stress of the precipitation hardened crystal.

Stage II. $r_f < D \leqslant r_s$: dislocations pile up at the grain boundary under static stress. Cross-slip out of one slip plane becomes more likely than in Stage I.

Stage III. $D \leqslant r_f < r_s$: as r_s is larger than r_f, dislocations are nucleated in two or three grains ahead of the moving crack. The size of the pile-ups is limited by the grain size, and under this condition the pile-ups will move in the opposite direction at the minimum stress per cycle.

Additional cross-slip and nucleation of slip at grain boundaries occurs as the zone size undergoes the transition from stage II to stage III, and this implies that the strain becomes more homogeneously distributed. The alloy will therefore approach the dislocation distribution observed in the overaged microstructure.

In stage I, the number of dislocations n_0 in the slip planes that produce slip steps at the crack tip can be calculated approximately by assuming that the group is extended to r_s,

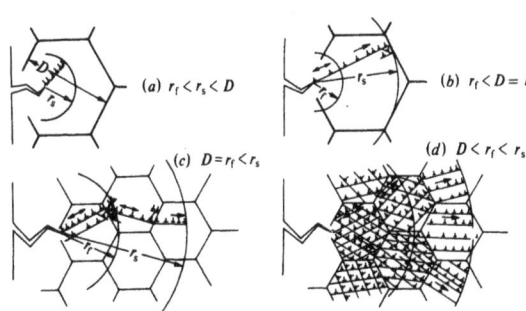

Fig. 3.34. Schematic representation of the dislocation arrangements if the plastic zone size varies, in comparison with the grain size. (After Hornbogen & Zum Gahr, 1976.)

$$n_0 = \Delta\tau_0 \, r_s/G \, b, \tag{3.38}$$

where $\Delta\tau_0$ is the critical shear stress of the precipitation-hardened crystal in which no particles have been sheared. If all n_0 dislocations contribute to crack propagation, the growth velocity is given by

$$da/dN = Cn_0 b, \tag{3.39}$$

where C is a dimensionless parameter which takes into account the orientation of the slip plane to the gross direction of crack propagation. Substitution of (3.26) gives

$$da/dN = C'(\Delta\tau_0/G) \, (K_{max}/p_Y)^2$$

$$= C'' K_{max}^2/p_Y E. \tag{3.40}$$

If it is assumed that $\Delta\tau_0$ is proportional to p_Y, all dimensionless parameters are included in C' or C''.

Considering next the behaviour of the overaged alloy (which exhibited homogeneous slip and a higher rate of da/dN for a given ΔK), Hornbogen & Zum Gahr point out that, unlike the aged and underaged alloys, partial reversal of primary slip activity will be unlikely. This is shown schematically in fig. 3.35, which represents crack propagation during one cycle in: (*a*) an overaged alloy, in which there is a tendency for homogeneous deformation, several slip planes and two slip directions are shown to be operating; (*b*) aged and underaged alloys, where the microstructure has a tendency towards inhomogeneous deformation, here the slip in one plane and system allows partial reversal of the primary crack propagation event. If, say, n_R dislocations move backwards on exactly the same slip plane, they will not contribute to the progress of the crack, and (3.39) becomes

$$da/dN = C \, (n_0 - n_R)b. \tag{3.41}$$

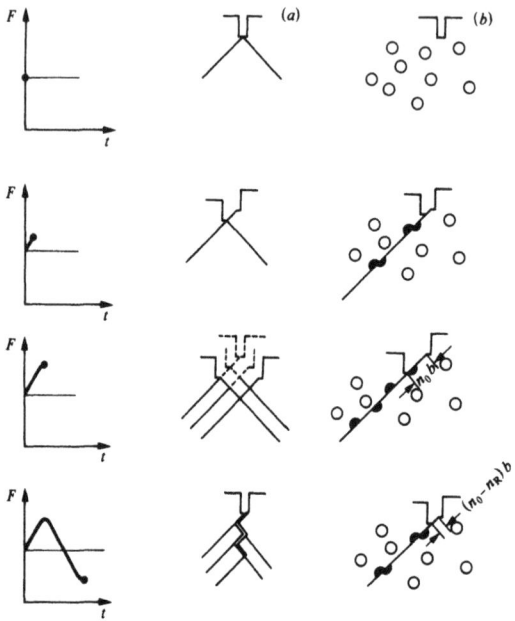

Fig. 3.35. Schematic representation of crack propagation during one cycle in: (*a*) an overaged alloy, the microstructure tending towards homogeneous deformation; and (*b*) aged and underaged alloys, the microstructure tending towards inhomogeneous deformation. (After Hornbogen & Zum Gahr, 1976.)

This model therefore accounts for the observed reduced crack growth velocity in the material in which heterogeneous slip processes were obtained. The process illustrated in figs. 3.34*a, b, c* and *d*, indicate that above the stress intensities at which the size of the static and the cyclic plastic zones reach the grain size, the number of reversible dislocations n_R will decrease, due to cross-slip at the grain boundary pile-ups. A larger proportion of the dislocations thus contributes to crack growth, which accounts for the observed sudden change in crack velocity.

Grain-boundary PFZs and incoherent second-phase particles can accelerate the rate of fatigue-crack growth, by interface decohesion. This effect is illustrated in fig. 3.36, which shows the crack propagation in vacuum as a function of stress intensity in the two alloys 7075 and X-7075 whose S–N curves were given in fig. 3.31. We saw that the presence of a dispersoid increased the fatigue life because of its effect in homogenizing the slip. However, fig. 3.36 shows that the beneficial effect of the dispersoid was wholly in its delaying fatigue-crack *nucleation,* since crack *propagation* rates are seen to be higher in the commercial 7075 alloy. Microstructurally, voids were nucleated at the dispersoid particles within

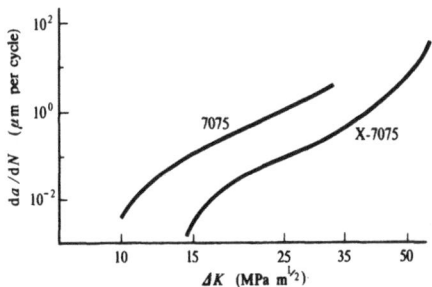

Fig. 3.36. Fatigue-crack growth curves for X-7075 alloy and commercial 7075 alloy for which the S-N curves are shown in fig. 3.31. (After Albrecht *et al.*, 1976.)

the plastic zone of the crack tip, and this accelerated the propagation process.

3.7.5   Possibilities for improving fatigue strength

The importance of achieving better fatigue properties has stimulated interest in the effects of thermomechanical treatments (TMT) on precipitation-hardened alloys. Deformation applied to an alloy in the solution-treated or partially-aged condition increases the dislocation density in the final aged state and produces a more uniform distribution of precipitate, so that fatigue deformation is more homogeneous and damage is less localized. Fig. 3.37 illustrates the effect of TMT upon a commercial Al-Zn-Mg-Cu alloy and also upon a high-purity alloy. Improved fatigue strength and endurance ratios were found, in spite of the presence in the former alloy of intermetallics and inclusions. This suggests that TMT may slow down early crack growth as well as influencing crack nucleation.

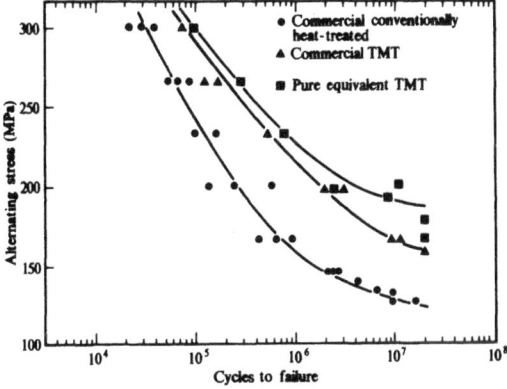

Fig. 3.37. S-N curves for Al-Zn-Mg-Cu alloys. (After Ostermann, 1971.)

More recently, Lu & Weissmann (1978) have compared the fatigue properties of an Al–14.4%wt% Zn alloy in the conventional aged condition and in the thermomechanically-treated condition, where each specimen possessed the same initial static yield stress. The TMT alloys exhibited, for high-cycle fatigue, a tenfold increase in fatigue life, and a 15% increase in the endurance limit. This improvement was attributed to an effective dispersal of slip, thereby increasing the resistance to crack nucleation.

Finally, it must be pointed out that there is inadequate fundamental knowledge about the optimum size and distribution of precipitates required for maximum fatigue strength. A mixture of small coherent particles to give strength, and larger dispersoids to homogenize strain distribution, may be the most effective, that is, if one can assume that no local weaknesses develop in PFZs or at particle/matrix interfaces.

4 Micromechanisms at elevated temperatures

4.1 Recrystallization of two-phase materials

We will consider separately the behaviour of alloys containing *fine* dispersions of a second phase, and alloys in which the phases are present in almost equal proportions as *coarse* grains.

4.1.1 Coarse two-phase alloys

Interest in this type of system appears to have re-awakened recently after a gap of over twenty-five years. For example, Mäder & Hornbogen (1974) have made a detailed study of deformed α/β brass.

The situation is a complex one, and the authors state that this alloy might be expected to show all phenomena which *can* occur in such alloys since the two phases differ strongly in crystal structure, diffusivity, stacking-fault energy and atomic order. Their results are summarized in fig 4.1.

For strains less than 40%, and annealing at the temperature of original precipitation of the α-phase (fig. 4.1a), the α-phase recrystallizes discontinuously from the grain-boundary sites while the β-phase shows 'continuous recrystallization' by subgrain growth. For annealing temperatures different from the original precipitating temperature, the volume fractions of the α- and β-phases must change so that α-nuclei may

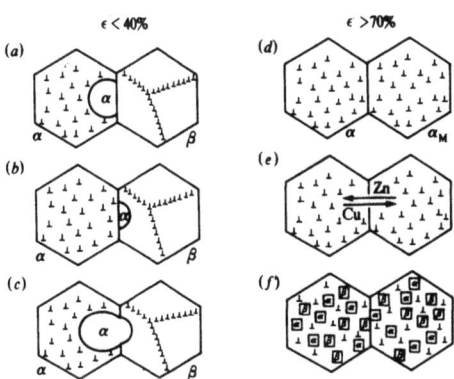

Fig. 4.1. Schematic representation of the micromechanisms of recrystallization in α/β brass for two different deformation strains. (After Mäder & Hornbogen, 1974.)

form within the β-grains (fig. 4.1*b*) which in turn act as recrystallization nuclei for the original deformed α-grains (fig. 4.1*c*).

For strains above 40%, the β-phase starts to transform martensitically into the α-phase (α_M), and this process is complete for strains above 70% (fig. 4.1*d*). On annealing (fig. 4.1*e*), cross-diffusion tends to equalize the compositions of both fcc phases, a fine-grained duplex structure forms (fig. 4.1*f*) via recrystallization nuclei of α-phase and retransformation nuclei of the β-phase.

In an investigation of this field by Vasudevan, Petrovic & Roberson (1974), a contrasting approach has been made. A model system (Ag-Ni) has been studied in which both phases are ductile and their compositions and volume fractions do not vary with temperature. Powder-metallurgically prepared samples containing 57 vol% Ag-phase and 43 vol% Ni-phase were compared with specimens of pure Ag and pure Ni.

Isochronal annealing experiments were conducted, and recrystallization was followed by macrohardness measurements on the alloy. Microhardness measurements were also conducted on the individual phases, and the results are shown in fig. 4.2. The dotted line represents the volume fraction hardness calculated from the measurements on the individual phases, and it is seen that this provides a good description of both shape and value of the experimental Ag-Ni macrohardness curve. The authors

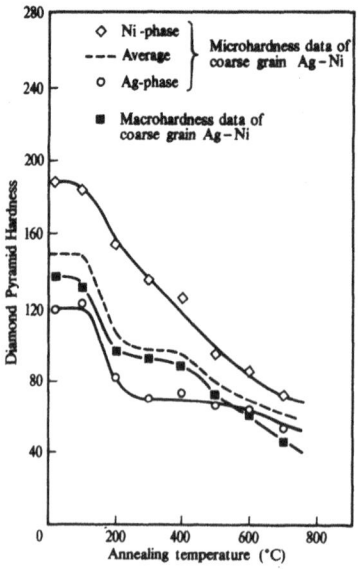

Fig. 4.2. Comparison of Ag-Ni macrohardness data to microhardness of the individual phases. (After Vasudevan, Petrovic & Roberson, 1974.)

reasonably conclude that the overall recrystallization is related to individual phase recrystallization on a volume fraction average basis.

4.1.2   Finely-dispersed two-phase alloys (incoherent particles)

Recrystallization kinetics

It is well established that dispersed, hard incoherent particles can either retard or accelerate recrystallization of a metallic matrix. Köster (1974) quotes a number of examples of this, of which figs. 4.3 and 4.4 are typical examples for the case of $CuAl_2$ in Al. These figures illustrate:

(i) the importance of interparticle spacing (λ) in determining behaviour, defined as the nearest-neighbour particle centre-to-centre distance;

(ii) that changes in *nucleation rate* with λ determine the overall rate of recrystallization (fig. 4.4), (the values of nucleation rate, $\hat{N}$, whose logarithms are quoted in fig. 4.4, are the apparent values, obtained by counting the number of grains when recrystallization is complete).

Particle size is important, in that acceleration of recrystallization is only observed in specimens containing relatively *coarse* particles (i.e. of diameter greater than 0.1 μm), when widely spaced. Very fine particles appear to give rise to retardation at all spacings, and the situation is summarized in fig. 4.5.

Finally, in this survey of the phenomenology, mention must also be made of the effect of the *degree of deformation* upon recrystallization kinetics in the presence of particles. Hansen & Bay (1972) have shown (fig. 4.6) in aluminium in which recrystallization is *retarded* by alumina dispersions, that the degree of retardation decreases as the degree of cold deformation is increased. It is suggested that the latter may reduce the

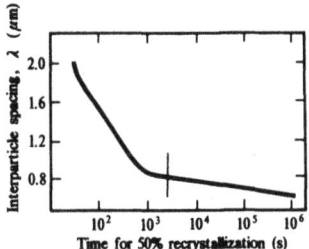 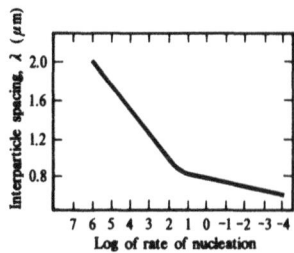

Fig. 4.3. Variation of interparticle spacing with time for 50% recrystallization in Al–$CuAl_2$ alloys. The time for the single-phase alloy is shown by the vertical line. (After Doherty & Martin, 1962-3.)

Fig. 4.4. Variation of the apparent rate of nucleation of recrystallization with the interparticle spacing in Al–$CuAl_2$ alloys. (After Doherty & Martin, 1962-3.)

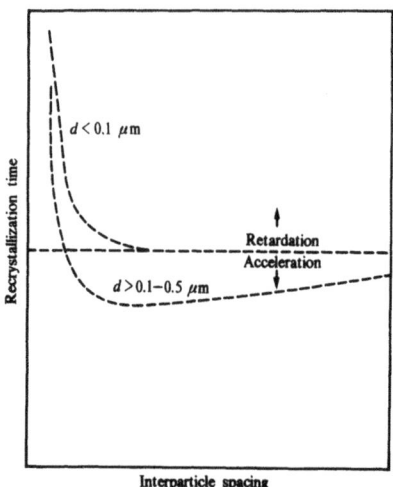

Fig. 4.5. The influence of particle size and particle spacing upon time of recrystallization of two-phase alloys. (After Hansen, 1975.)

critical size of the recrystallization nuclei, so that by increasing the cold deformation, the critical size of the nucleus can be reduced to be of the same order, or smaller than the particle spacing: the dispersion would then only affect the *growth* of the nuclei. Measurements of nucleus sizes and growth rates in dispersion-hardened products appears to be desirable in this context.

Micromechanisms

Retardation effects. Hard second phases will tend to increase the stored energy of the deformed material through their effect of promoting dislocation multiplication during cold work, as discussed in chapter 2. Retardation of recrystallization can arise through two effects. Firstly, the second-phase particles may cause an inherently stable dislocation structure. Secondly, the particles may hinder dislocation (and network) rearrangements thus hindering the formation of grain boundaries and their migration.

As discussed earlier (chapter 2), finely-dispersed small particles have the effect of homogenizing the distribution of slip, so that the deformation structure contains no regions of severe plastic curvature and is thus inherently more resistant to recrystallization. This effect has been convincingly demonstrated in single crystal specimens deformed to modest strains. For example, Rollason & Martin (1970) took a crystal of Cu-2wt% Co which had been aged to generate a dispersion of incoherent Co particles, and deformed it into stage II of the stress-strain curve. It

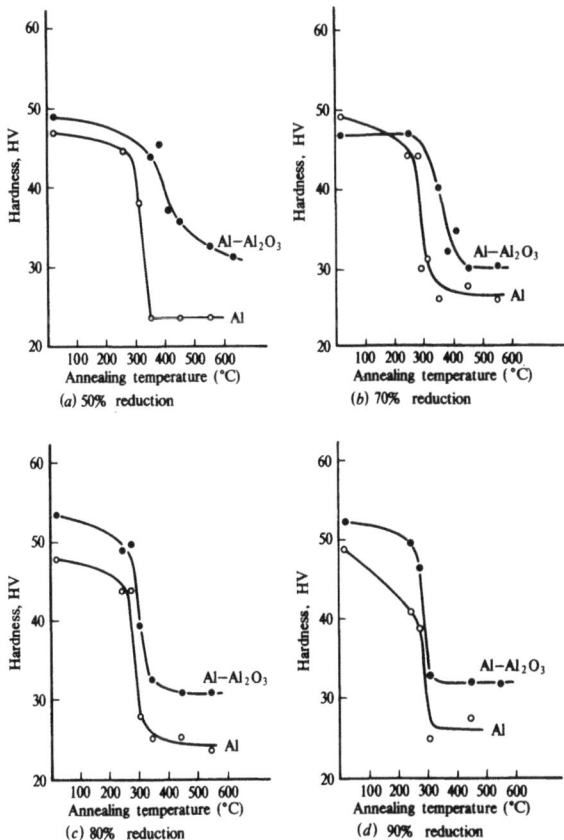

Fig. 4.6. The influence of the degree of prior strain upon the recrystalization kinetics of Al-Al₂O₃. Curves of hardness versus annealing temperature for : (*a*) 50% reduction; (*b*) 70% reduction; (*c*) 80% reduction; and (*d*) 90% reduction. (After Hansen & Bay, 1972.)

was then annealed for 98 hours at 950 °C, at which temperature all the Co particles would dissolve within a few seconds, yet it was seen that no recrystallization occurred. It is unlikely that this effect would be important in polycrystalline materials, however, due to the complex nature of the slip occurring. This would lead to nonlaminar slip and this to high inhomogeneities of deformation. Therefore, any high resistance to recrystallization in polycrystalline materials must arise from *particle pinning* effects and not from grain boundaries or dislocation sub-boundaries. Particle pinning effects can be divided into two main categories.

(*a*) Grain-boundary pinning. Zener (quoted by Smith, 1948) was the first to point out that a dispersed phase will retard a grain boundary under a fixed driving force. As discussed in §2.1.4, when a migrating boundary of specific surface energy σ_B intersects a spherical inclusion of radius r, the maximum value of the pinning force exerted by the particle is given by $\pi r \sigma_B$. If there are N of these particles per unit volume, their volume fraction f is $\frac{4}{3}\pi r^3 N$. A boundary of unit area will intersect all particles within a volume $2r$, i.e. $2rN$ particles, so the number of particles intersecting unit area of a boundary is given by $3f/2\pi r^2$. Thus the retarding force per unit area of grain boundary (F_r) is given by

$$F_r = 3f\sigma_B/2r. \tag{4.1}$$

(b) Sub-boundary pinning. The maximum retarding force acting on a particle in a sub-boundary is given by $\pi r\,\sigma_{sb}$, where σ_{sb} is the specific sub-boundary surface energy. If all the particles lie on subgrain boundaries, then the number of such particles per unit area of subgrain boundary is given by (2.33) as

$$N_S = 3fL/8\pi r^3$$

where L is the subgrain size. The retarding force per unit area of subgrain boundary (F_{sr}) is thus given by

$$F_{sr} = 3fL\,\sigma_{sb}/8r^2. \tag{4.2}$$

Acceleration of recrystallization. This situation arises, as shown in fig. 4.5, above some critical particle size. Gawne & Higgins (1971) have carried out careful quantitative metallographic analyses which have convincingly demonstrated that, in an Fe-C alloy, carbide particles act as nucleation sites for recrystallization. They also showed that a critical particle size is necessary for this effect. More recently, Humphreys (1977) has studied the conditions under which dispersed particles can provide nuclei for recrystallization of deformed aluminium.

Humphreys' experimental material, already described in §2.2.1, consisted of single crystals of aluminium containing 0.45wt% Cu and varying amounts of silicon, aged to produce equiaxed dispersions of Si particles in an Al-0.45wt% Cu matrix. Particle sizes in the range 0.7 to 4.9 μm were chosen, and various cold-rolling reductions were applied to the crystals, after which they were annealed at 300 °C (at which temperature the dispersed phase was stable), and the regime of 'particle stimulated nucleation' was identified metallographically. Because of the large interparticle spacing, the association between particles and nuclei could be inferred directly, without the need for the detailed statistical analysis used by Gawne & Higgins, and, as shown in fig. 4.7, Humphreys was able clearly to distinguish between those specimens where a substantial number of particles were associated with recrystallization nuclei, and those

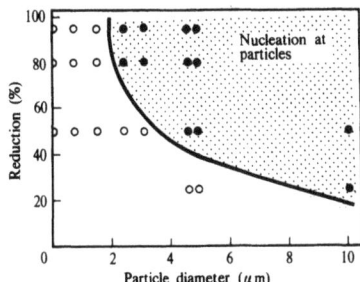

Fig. 4.7. The conditions of deformation and particle size for which nucleation is observed to occur at particles. Closed circles indicate observations of nucleation at particles. (After Humphreys, 1977.)

specimens where there was no such association.

It is seen that the critical particle size for stimulated nucleation in these crystals is 2 to 3 μm, and it appears that at least 25% strain is required to cause recrystallization. By direct observation of *in situ* annealing processes in the high voltage electron microscope, Humphreys has identified the processes by which a recrystallization nucleus may form in association with a second-phase particle under the conditions discussed above.

As discussed in §2.2.1, the as-deformed structure consisted of a set of subgrains whose size decreased adjacent to each particle, as shown diagrammatically in fig. 2.25*i*. Electron diffraction experiments indicated that cumulative misorientations of up to 15° developed as a particle was approached. Fig. 4.8 indicates diagramatically the observed annealing behaviour within the deformation zone adjacent to a particle: a nucleus A in the deformation zone grows at an early stage towards and around the particle. After the deformation zone has been consumed, the driving force for recrystallization arises from the dislocation density (which is considerably lower in the matrix outside the deformation zone), and so recrystallization cannot proceed unless the driving force can overcome the retarding force due to the grain boundary. There is thus a critical nucleus size, r_c, given by

$$r_c > 2\sigma_B/E,$$

where σ_B is the grain-boundary energy and E the stored energy. If r_c is taken to be equal to one-half the particle diameter, then this analysis leads to a critical particle diameter of the order of 1 μm (which decreases with increasing stored strain-energy, E).

The effect of slip distribution. Kamma & Hornbogen (1976) have examined the effect of carbide size upon the initiation of recrystallization in a hypo-eutectoid steel. They produced varying sizes of a fairly randomly

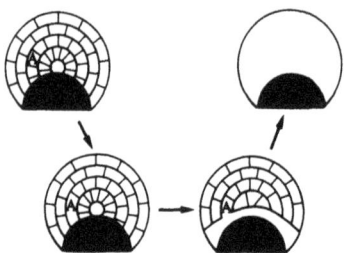

Fig. 4.8. The annealing behaviour of a deformation zone adjacent to a particle. (After Humphreys, 1977.)

dispersed Fe_3C particles by thermomechanical treatments of the martensitic structure. Three dispersions were studied, namely coarse spheres (2700 nm diameter), medium-sized spheres (200 nm diameter) and fine plates (200 × 10 nm), and we will refer to these three structures as C, M and F respectively.

Fig. 4.9 shows the time required to initiate recrystallization at a recrystallization temperature of 550 °C as a function of cold work for structures C, M and F. Two unexpected effects are apparent.

First, in the alloy (F) containing the very small particles, recrystallization starts earlier (over most of the range of ϵ) than in alloy M which contains larger particles. This disagrees with (4.1), while the early initiation in alloy C is in agreement with the model for acceleration of recrystallization described above.

Secondly, fig. 4.9 shows that the start of recrystallization in F is delayed at very high amounts of cold work to times longer than that of alloy M.

Although this work confirms that there is an upper critical particle size above which the initiation of recrystallization is accelerated, and that recrystallization may be retarded by smaller particles, a further variable needs to be considered. There appears to be a minimum critical particle size below which (in the case of alloy F) particles are sheared during deformation. This leads to microscopic heterogeneity of strain, as previously discussed (§3.2). Such strain concentration will *favour* the formation of nuclei for recrystallization. This accounts for the curve for alloy F lying to the left of that for alloy M in fig. 4.9. At higher strains (ϵ greater than 80%), the deformation bands will broaden and strain will become homogeneous again. Only then is recrystallization retarded still more in alloy F than in alloy M.

4.1.3   The effect of duplex dispersions on recrystallization

As pointed out at the start of chapter 1, many technologically important alloys contain a mixed particle structure, consisting of a coarse distribution

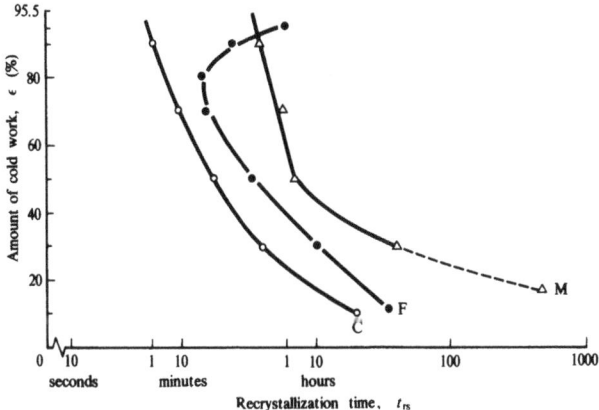

Fig. 4.9. The time to start recrystallization as a function of the amount of
cold work and pre-treatment of a hypo-eutectoid steel recrystallized at
550 °C. (After Kamma & Hornbogen, 1976.)

of large inclusions due to the casting, and a fine-particle dispersion introduced during subsequent processing. The recrystallization behaviour of
a material containing such a duplex structure has been considered
theoretically by Nes (1976a), and fig. 4.10 indicates the model used in
his approach. The system contains three principal structural elements –
a relatively coarse distribution of large inclusions, a finer dispersion of
smaller particles, and a subgrain structure due to cold working.

Nes assumes that nucleation of new grains occurs very rapidly at the
large inclusions, which is termed a 'site-saturated' reaction, i.e. a case
where all grain nucleation events take place within a short initial transformation period during which only a small volume fraction has
recrystallized. The final grain size will thus be determined by the number
of sites, and be independent of both the nucleation and growth rate of the
grains. The performance of the coarse particles as nucleation sites will be
drastically affected by varying the level of the surrounding fine-particle
dispersion. Nes assumes that the particles would restrict subgrain growth
to subgrains of larger than average sizes, and that these subgrains only
contribute in the grain nucleation process.

By selecting these subgrains, it is concluded by analogy with the
analysis of abnormal grain growth due to Hillert (1965) that only
subgrains of diameters ρ given by the following relation will grow:

$$\rho > \frac{\bar{\rho}}{1 - \frac{3}{4}\alpha\bar{\rho}f/r} , \qquad (4.3)$$

where $\bar{\rho}$ is the average subgrain diameter, r is the particle radius, f their
volume fraction and α a constant of order unity.

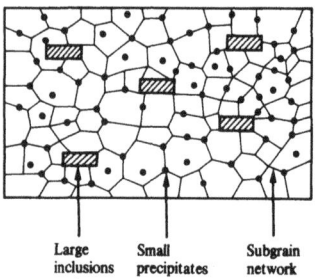

Large   Small   Subgrain
inclusions precipitates   network

Fig. 4.10. Model of duplex dispersion employed by Nes (1976*a*).

The probability for a nucleation event in association with a possible site is proportional to the number of subgrains able to grow. If n is the number of subgrains able to grow per unit volume and $\bar{n}$ the average subgrain density, then N, the number of successful nucleation events per unit volume, will be

$$N = N_0 \, n/\bar{n}, \tag{4.4}$$

where N_0 is the average density of possible nucleation sites in the matrix. The number, n, of subgrains able to grow is given by the integral

$$n = \int_{\rho}^{\infty} f(\rho) \, \mathrm{d}\rho, \tag{4.5}$$

where $f(\rho)$ is the subgrain size distribution function, and ρ is given by (4.3). By assuming a Gaussian subgrain distribution, this integral gives

$$n = \bar{n} \, [1 - \mathrm{erf}\,(\tfrac{1}{2}\,(\rho - \bar{\rho})/\sigma)], \tag{4.6}$$

where σ is the standard deviation in the subgrain distribution. By combining (4.3), (4.4) and (4.6), Nes obtains

$$N = N_0 \left[1 - \left(\frac{\bar{\rho}^2}{\sigma\sqrt{2(4r/3f\alpha) - \bar{\rho}]}} \right) \right], \tag{4.7}$$

which relates the number of successful nucleation events to the number of heterogeneities, N_0, the subgrain distribution, σ and ρ, and the particle dispersion characteristics, f/r. Equation (4.7) thus describes the nucleation rate, since under the site-saturated conditions described, $\dot{N}/\dot{N}_0 \equiv N/N_0$.

In an investigation of the recrystallization behaviour of an Al–Mg–Si alloy, Scharf & Gruhl (1969) have quantitatively determined the variation in nucleation rate with the particle dispersion level, f/r, and their results are shown in fig. 4.11, together with the theoretical curve which is derived according to (4.7) using $\alpha = 0.6$, and estimating values of subgrain distri-

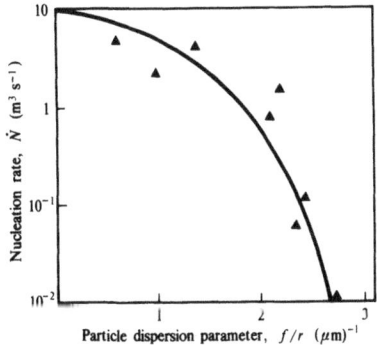

Fig. 4.11. Recrystallization data of an Al–Mg–Si alloy, due to Scharf & Gruhl (1969), plotted according to model of Nes (1976*a*). The full curve is derived from (4.7) using $\alpha = 0.6$. (After Nes, 1976*b*.)

bution parameters $\bar{\rho}$ and σ. It is seen that the experimental results are well accounted for by the analysis of Nes (1976*a*). The effects of varying the dispersion levels on the nucleation rates for the situations when r is greater or less than r_{crit} (r_{crit} being the critical size above which the particle may act as a nucleation site) are shown in fig. 4.12.

Fig. 4.12 expresses the same information as fig. 4.5, but employs the parameter f/r (which is proportional to the pinning force) instead of the interparticle spacing (λ). In fact $1/\lambda$ is proportional to $f^{\frac{1}{3}}/r$, so that an approximate reciprocal relationship is to be expected between the two diagrams.

This illustration thus summarizes the effects of dispersed particles upon the kinetics of recrystallization. We will next consider the effects of elevated temperatures upon the response of particle-hardened materials to applied stresses.

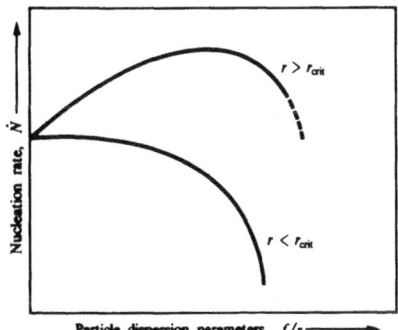

Fig. 4.12. The variation of nucleation rate of recrystallization with f/r according to the model of Nes (1976*a*).

4.2   Changes in the yield stress

It is convenient to study model systems consisting of single crystals hardened with oxide particles when attempting to understand the interaction of dislocations with particles as a function of temperature, because the distribution of oxide particles remains stable at high temperatures. There has been considerable work on the high-temperature strength of internally oxidized copper crystals, and fig. 4.13, which refers to such crystals containing various volume fractions of spherical particles of SiO_2, is typical of the results obtained.

In fig. 4.13, the shear yield stress/shear modulus ratio is plotted as a function of test temperature between 77 K and 1200 K. At low temperatures, the curves are horizontal, indicating that in this range the variation of the Orowan stress with temperature is mainly due to changes in the matrix shear modulus (2.28). However, as the temperature rises a substantial fall in yield stress occurs, and this is attributed to climb of dislocations around the particles. It is significant that these crystals retain a large fraction of their strength at temperatures that are close to the melting point of the copper matrix. The minimum flow stress is found to

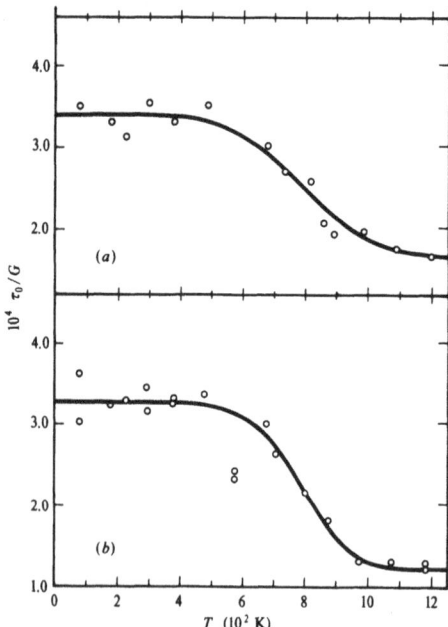

Fig. 4.13. The modulus-corrected yield stress as a function of testing temperature for: (a) Cu-O.41 vol.% SiO_2; and (b) Cu-1.25 vol.% SiO_2. (Courtesy of G. L. Lloyd.)

be approximately one-half of the Orowan stress observed at low temperature.

Three principal theories have been proposed to account for the above effects.

The theory of Brown & Ham (1971). This model, developed to explain the higher-temperature strength of particle-hardened crystals, assumed that during plastic deformation the dislocations remain in their slip planes except that the yield stress was controlled by the 'local' climb of dislocations over the particles. They realized that when a dislocation does climb locally over a particle it must increase its line length. This requires energy and explains why the yield stress of dispersion-hardened crystals does not become vanishingly small as the temperature approaches the melting-point of the matrix. Brown & Ham calculated the yield stress when all the particles are bypassed by local climb, and arrived at a yield stress of one-half the Orowan stress, which agrees quite well with experiment.

In the theory, Brown & Ham assumed that the chemical potential for the local climb of a dislocation over a particle was large compared to kT, where k is Boltzmann's constant and T the temperature. Their theory predicts a linear decrease in τ with temperature. Their final expression is:

$$\tau = \tfrac{1}{2}\frac{Gb}{(1-\nu)^{\frac{1}{2}}L} + \frac{SU_p}{\sqrt{2}Lb^3} + \frac{SkT}{\sqrt{2}Lb^3}\ln\left(\frac{\dot{\epsilon}S^2}{2\rho b^4 V_D}\right), \qquad (4.8)$$

where

$\quad U_p$ is the activation energy for dislocation 'pipe' diffusion,
$\quad \rho$ is the dislocation density,
$\quad \dot{\epsilon}$ is the strain-rate,
$\quad V_D$ is the Debye frequency,
$\quad \nu$ is the Poisson's ratio,
$\quad S$ is the side length of the particle (assumed to be cubic),
$\quad L$ is the interparticle spacing.

Qualitatively, the theory agrees with the experimental data available, however, the predicted slope of the τ/G against temperature curve is typically an order of magnitude higher than that actually observed. Humphreys, Hirsch & Gould (1970) have pointed out that the chemical potential at yield is actually small compared to kT. If (4.8) is re-derived on this assumption, then the stress is directly proportional to the strain-rate, instead of logarithmically as observed.

The theory of Humphreys, Hirsch & Gould (1970). These authors have suggested that the nucleation of jogs may be an important factor in these alloys. By comparing the rates of jog nucleation and of jog migration, they conclude that dislocation climb at the particles is controlled by *jog nucleation*, the process being as illustrated in fig. 4.14, which shows an edge dislocation climbing over a spherical particle. The segment ABC has climbed out of the slip plane DCE, by the generation and diffusion of jogs, enabling the dislocation to advance, which provides the driving force for the process. Their final equation describing the climb of dislocation is

$$[cG(\tau - \tau_b)]^{\frac{1}{2}} = \frac{U_p}{v} + \frac{2W_{jo}}{v} - \frac{kT}{v} \ln \left(\frac{4A\ V_D(\tau - \tau_b)\ b^5\ \rho D^2}{d^2\ \dot{\epsilon}\ kT} \right),$$

(4.9)

where

$$v = \tfrac{1}{2}(D/d)^{\frac{1}{2}}\ b^3,$$

and

τ_b is the back-stress due to the dislocation increasing its line
length when climbing over a particle,
W_{jo} is the activation energy for jog formation,
A is an entropy factor,
D is the interparticle spacing,
d is the effective particle spacing in the slip plane,
c is a constant ($c = 1$ for edges, $c = (1 + v)/(1 - v)$ for screws).

This expression accounts for the observed logarithmic dependence of the yield stress with strain-rate, and the available experimental data, when plotted in terms of this relationship, provides a value of the activation energy for climb in good accord with the theoretical value of ($2W_{jo} + U_p$). Nevertheless, the theoretical slope of the logarithmic dependence of the yield stress with strain-rate is a factor of 10 less than the experimental slope.

The theory of Shewfelt & Brown (1977). These authors have made calculations of the velocity of an edge dislocation passing through a random array of point obstacles, some of which are bypassed by local climb. The results of the calculations were compared with the results of hot tensile tests on copper crystals containing spherical SiO_2 particles.

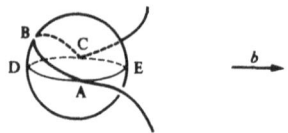

Fig. 4.14. An edge dislocation climbing over a spherical particle.

The most important result of the calculations is that the dislocation velocity was found to be exponentially dependent on the applied shear stress (even though the time required for the dislocation to overcome an individual obstacle depends only weakly on the stress). This agrees with the experimental results that the yield stress varies logarithmically with the strain-rate at constant temperature. The experimental data (for the particle dispersions considered) were fitted to the following equation:

$$\tau = Gb\,[(0.51 + 0.01) + (0.12 \pm 0.02)\log{(\dot{\varepsilon}kTR^2/4\pi\rho b^2 a_v GDD_S)}$$
$$+ (0.052 \pm 0.009\,E_s/kT]/D \qquad (4.10)$$

where

$$D_s = D_0 \exp{(-E_s/kT)},$$

and

ρ is the dislocation density,
R is the particle radius,
D is the square lattice spacing of the obstacles,
a_v is the area associated with a vacancy.

Reasonable agreement with experiment was obtained by these authors, who conclude that no new mechanism, such as the jog mechanism discussed earlier, need be postulated to explain the exponential dependence of strain-rate with the applied stress. The theory also predicts the limiting values of the climb-controlled yield stress at high temperatures to be 0.4 of the Orowan stress, which agrees well with the experimental results. It may be concluded that refinements of computer approaches may eventually lead to further improvement in the degree of agreement between theory and experiment in this area.

4.3   Changes in the rate of work-hardening

Work on model systems again provides the clearest evidence of the changes in work-hardening rate with increasing temperature, and electron metallographic study of deformed single crystals has, furthermore, enabled an understanding to be gained of the micromechanisms of deformation. We have already mentioned the use of copper-based crystals containing non-deformable (usually oxide) precipitates, whose deformation behaviour has been analysed in detail in recent years. In the present context we will take as our example an aluminium-based alloy. Most commercial aluminium alloys contain plate-like precipitates on ageing, and such structures are unlikely to behave in a simple manner on deformation (see chapter 2), and in order to simplify the quantitative evaluation of the effects of the precipitate, uniformly dispersed *equiaxed* particles are desirable.

Stewart & Martin (1975) have studied single crystals of Al-0.5wt% Si-0.45wt% Cu aged to produce equiaxed silicon particles within a solution-hardened matrix (see §1.3.3). We will consider first their deformation behaviour, second the micromechanisms operating, and then go on to discuss the theoretical interpretation of the various processes.

Fig. 4.15 illustrates the changes in the form of the shear stress–shear strain curves observed as the testing temperature is increased from 77 K to 493 K. The crystals contained a volume fraction of 0.31% of Si particles of mean planar diameter (d_S) of 0.097 μm with a mean planar interparticle spacing (λ_S) of 1 μm. The curves show a parabolic region up to strains of 30%, and the orientation of the crystals was such that this range of deformation involved essentially primary slip activity. Fig. 4.16 shows the temperature dependence of the parabolic work-hardening rate $(d\tau/d\epsilon^{\frac{1}{2}})$. A very high temperature sensitivity is apparent, especially in the temperature region 300 to 500 K.

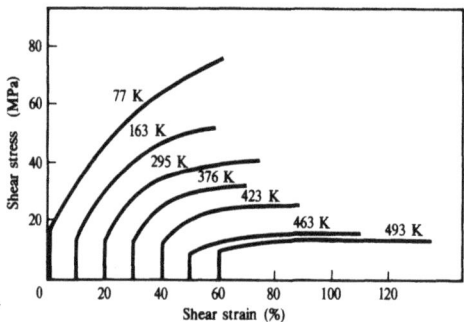

Fig. 4.15. Shear stress–shear strain curves for aluminium crystals containing 0.31 vol.% of Si particles, tested between 77 K and 493 K. (After Stewart & Martin, 1975.)

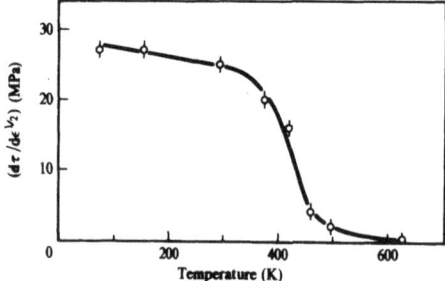

Fig. 4.16. The temperature dependence of the parabolic work-hardening rate of the crystals whose stress–strain curves are shown in fig. 4.15. (After Stewart & Martin, 1975.)

The dislocation structures produced on deformation at temperatures in the range of 77 to 49 K have also been investigated. Considering a standard shear strain of 6% for comparison purposes, the dislocation arrays produced between 77 K and room temperature were essentially as described in chapter 2 (§2.2.1), and illustrated in fig. 2.20. The structure in this temperature range consists of rows of primary prismatic loops aligned with the particles in the direction of the primary Burgers vector. As in the case of copper single crystals containing dispersed oxide particles, the dislocation arrays generated in the vicinity of the particles were found to be a function of the particle size (as indicated in fig. 2.24a), the arrays becoming more complex as the particle size increased.

The crystals were orientated for (111) [$\bar{1}$01] glide, and table 4.1 summarizes the dislocation arrays generated, the Burgers vectors involved, and the approximate critical particle sizes at shear strain of 6% where the arrays first appeared.

Marked changes in primary dislocation structure were observed between specimens deformed at room temperature and specimens deformed at elevated temperatures, especially in the temperature range where there is a large reduction in the work-hardening rate (fig. 4.16). In this range, the number of prismatic loops decreases, so that after deformation at 425 K, aligned loops are only found in association with larger particles. At 463 K, as shown in fig. 4.17, there is a large reduction in primary dislocation density: primary prismatic loops and helices are no longer observed, and few dipoles are seen in association with the particles.

Additional factors affect the formation of dislocations with secondary Burgers vectors. As the temperature of deformation increases above room temperature, the critical particle size for secondary dislocation generation increases markedly, so that at 423 K it has risen to 0.26 μm. A similar pattern of behaviour has recently been reported by Humphreys & Hirsch in copper single crystals containing small particles of SiO_2, namely that the number of both primary and secondary dislocation loops associated with the particles drops significantly with increasing temperature of deformation.

Table 4.1. *The basic dislocation structures in the crystals deformed at room temperature to a shear strain of 6%.*

Structure	Critical particle size (μm)	Burgers vectors present
Primary	<0.12	[$\bar{1}$01]
	≈0.12	[$\bar{1}$01] [110]
Primary + secondary	≈0.18	[$\bar{1}$01] [110] [011]
	≈0.21	[$\bar{1}$01] [110] [011] [101]

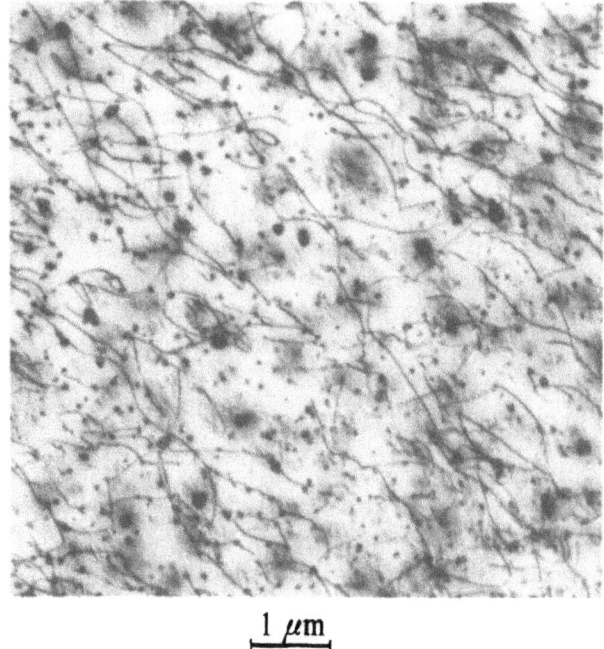

$\underset{\longmapsto}{1\ \mu\text{m}}$

Fig. 4.17. Transmission electron micrograph of (111) section of a single crystal of aluminium containing a dispersion of silicon particles plastically deformed 6% at a temperature of 463 K. (Courtesy of A. T. Stewart.)

This pattern of behaviour is consistent with a model proposed by Gould, Hirsch & Humphreys (1974). Above a certain temperature depending on the strain-rate, the particle size and the position of the loop relative to the equatorial plane of the particle, Orowan loops shrink and disappear by climbing around the particles. Fig. 4.18 illustrates the mechanism whereby an Orowan loop may climb out by transfer of vacancies from jogs AA' to jogs BB', by diffusion along the dislocation pipe.

The consequent decrease in internal stress around the particle reduces the chance of formation of prismatic loops by double cross-slip (see §2.2.1). Thus, both the number of Orowan loops and the number of primary prismatic loops will decrease with increasing temperature, resulting in the observed decrease in work-hardening. The decrease in the number of Orowan loops by climb with increasing temperature in the recovery range will also lead to a decrease in the number of secondary loops generated, or, at a given strain, to an increase in the size of particle at which such loops are generated.

Hirsch & Humphreys (1969) derive an expression for the critical temperature, T_c, at which the work-hardening disappears by the above

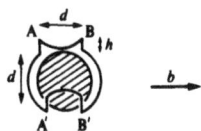

Fig. 4.18. Climb of an Orowan dislocation loop at a particle; cross-slip has occurred, and the loop can anneal out by climb by vacancy transfer from jogs AA' and BB' along the dislocation pipe. As it climbs, the loop shrinks around the particle due (1) to its line tension and (2) to the attraction of the screw segments AB, A'B'.

processes. They firstly calculate the time, t, for a loop to disappear by the process illustrated in fig. 4.18. Assuming the rate of climb to be diffusion-controlled, the flux of vacancies is

$$\phi = - \frac{D_p}{Vc_0} \text{ grad } c, \tag{4.11}$$

where

D_p is the diffusion coefficient for pipe diffusion,

c_0 is the equilibrium concentration of the vacancies,

c is the actual vacancy concentration,

V is the atomic volume ($V = b^3/\sqrt{2}$).

The climb force F_{cl} on jogs A and B is equal and opposite, so that, using a theory of climb due to Friedel (1964), they write

$$\phi = - \frac{zA\nu_0 b^2 \exp\left(-U_p/kT\right)}{Vd} \left[\exp\left(\frac{F_{cl}b^2}{kT}\right) - \exp\left(\frac{-F_{cl}b^2}{kT}\right)\right], \tag{4.12}$$

where

h,d are height and separation of jogs as shown in fig. 4.18,

z is the coordination number (≈ 11),

A is an entropy factor (≈ 10),

ν_0 is the Debye frequency ($\sim 10^{13}$ s^{-1}),

U_p is the activation energy for pipe diffusion,

k is the Boltzmann factor,

T is the temperature.

The total flux of vacancies is $2\phi a$, where a is the cross-section of the pipe. The number of vacancies to be transported is approximately dhb/V, so that the time t taken for a loop to disappear is given by

$$1/t = 2\phi aV/dhb. \tag{4.13}$$

Substituting (4.12) and making the reasonable assumption that $F_{cl}b^2/kT$ is small, they obtain

$$1/t = (4azA\nu_0 b/d^2 h)\,(F_\text{cl}b^2/kT)\exp{(-U_\text{p}/kT)}. \tag{4.14}$$

The number of loops produced per second per particle is given by differentiation of an equation of the form of (2.50). Hirsch & Humphreys write

$$dN/dt = 2r\dot{\epsilon}/b, \tag{4.15}$$

where r is the particle radius and $\dot{\epsilon}$ the plastic strain-rate.

The hardening effect from the loops will thus disappear if

$$1/t \leqslant dN/dt,$$

since no loops are formed. This gives T_c from (4.14) as

$$T_\text{c} = \frac{U_\text{p}}{k \ln{[(2azA\nu_0 b^2/d^2 hr\,\dot{\epsilon})\,(F_\text{cl}b^2/kT_\text{c})]}}. \tag{4.16}$$

The most important contribution to the climb force, F_cl, is taken to be the dislocation line tension, for as the loop climbs it shrinks round the particle. This effect gives

$$F_\text{cl} \approx \frac{\alpha Gb^2 \ln{(d/r_0)}}{2\pi d}, \tag{4.17}$$

where r_0 is the effective core radius, and α depends on the height of the loop relative to the centre plane of the particle. For an average position of the loop they take α as approximately $\frac{1}{2}$. Writing the pipe cross-section $a = f^*\pi b^2$, combination of (4.16) and (4.17) give

$$T_\text{c} = \frac{U_\text{p}}{k \ln{[(f^*zA\nu_0 b^5 \alpha/d^3 hr\,\dot{\epsilon})\,(Gb^3 \ln{(d/r_0)}/kT_\text{c})]}}. \tag{4.18}$$

For a pipe radius of approximately $3b$, $f^* = 10$.

Referring to the experimental data shown in fig. 4.16, and taking $T_\text{c} = 420$ K, it is found that U_p is about 0.85 eV, which is not unreasonable value.

The above theory appears, therefore, to account reasonably well for the observed experimental observations upon single crystal specimens, namely the dramatic loss of work-hardening encountered at a relatively low temperature.

4.4   Creep in particle-hardened alloys

At low temperatures materials deform by dislocation glide, and as the temperature increases other mechanisms of flow become possible. Dislocations can climb as well as glide, grains slide over each other at grain boundaries, vacancies diffuse. Thus, in a polycrystalline solid there are many distinguishable mechanisms by which it can flow. In one range of stress and temperature one of these flow mechanisms is dominant, in another range a different mechanism will obtain.

According to Weertman & Weertman (1965) and Ashby (1973) this 'landscape' of deformation mechanisms is most conveniently surveyed with the aid of a 'deformation mechanism map'. Fig. 4.19 is such a map for pure nickel with a grain size of 100 μm. It shows the fields in which a given flow mechanism is dominant, and the strain-rate due to all the mechanisms acting together. The coordinate axes are temperature, T, normalized in respect to the melting point, T_m, and normalized shear stress, τ/G, on a logarithmic scale.

At low temperatures flow is confined to the *dislocation glide* field, slip being the dominant mechanism. Above $0.3T_m$ dislocations can climb as well as glide and the flow has different characteristics known as *power-law creep*. In metals there are two subgroups of power-law creep: at higher temperatures lattice diffusion is the dominant transport mechanism; but at lower temperatures dislocation core-diffusion becomes the more efficient process. Finally, at still higher temperatures *diffusional flow* by the

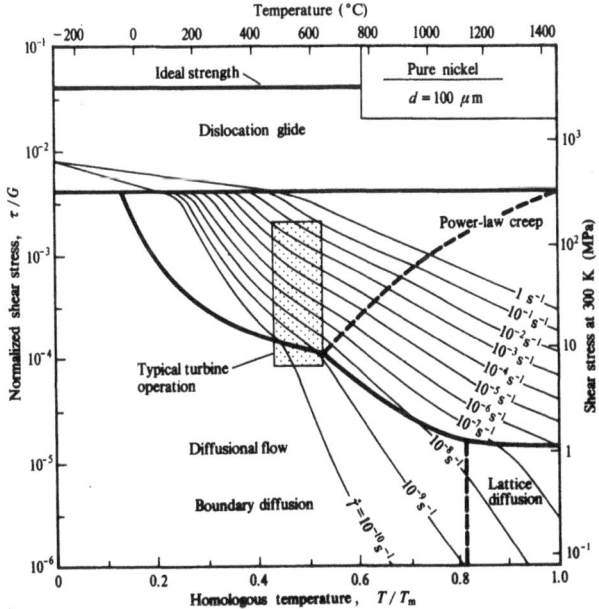

Fig. 4.19. The deformation mechanism map for pure nickel of grain size 100 μm. The shaded box indicates the typical stresses and temperatures to which a turbine blade is subjected. (After Ashby, 1973.)

motion of single ions permits deformation in a polycrystal at a rate which depends strongly on grain size. Like power-law creep, two subgroups exist, corresponding to diffusion occurring predominantly through the grains (Nabarro–Herring creep) or round their boundaries (Coble creep).

Diagrams like fig. 4.19 are constructed from a set of rate equations or constitutive laws, and data about the material which allows the equations to be evaluated. They are, of course, an oversimplification and should not be treated as complete or exact. They are limited to the description of *steady-state* flow, and this is the type of creep to which we will confine our attention in the present section. The steady-state creep rate for high temperatures and moderate stress (p) may be described by a 'power-law' equation of the form

$$\dot{\gamma} = \frac{A_s DGb}{kT} \left(\frac{p - p_0}{G}\right)^n, \tag{4.19}$$

where $\dot{\gamma}$ is the shear strain-rate, A_S and n are material constants, D is a diffusion coefficient and p_0 is the internal or threshold stress. Though p_0 is negligible in pure metals, the creep of certain dispersion-hardened alloys suggests the existence of such a threshold.

4.4.1   The suppression of power-law creep

Ashby (1973) points out that *intrinsically* high creep strength will be observed in materials in which diffusion is slow, and in which dislocations are poor sinks or sources for point defects. The first of these is achieved in covalent elements and compounds, and materials of high melting-point. The second is an attribute of a low stacking-fault energy, since the core attachment or detachment kinetics are slowed when dislocations are widely dissociated.

Further reduction in power-law creep may be achieved by metallurgical control designed to slow both glide and climb. Solute and particle hardening may reduce the former. Climb is slowed by alloying which lowers the diffusivity, lowers the stacking-fault energy and which introduces an internal or threshold stress. A grain-boundary precipitate will also have the effect of pegging the boundaries and thus suppressing grain-boundary sliding.

The most sophisticated family of alloys exemplifying the above principles are the age-hardening nickel-based 'superalloys' which are widely used in gas turbines. Improvements in the physical metallurgy and processing of the superalloys have made possible an average increase of 10 K per year in operating temperature over the last 35 years, a record unmatched by any other alloy development.

The basis of superalloys is a Ni–Fe–Cr austenite (γ). A high nickel content confers improved basic oxidation resistance, improved solubility of alloying elements and better phase stability. The chromium addition gives excellent resistance to oxidation and hot corrosion. An increased iron content may be made in order to reduce material cost. The principal hardening phase is γ'-Ni$_3$Al or Ni$_3$(Al,Ti) which almost exactly matches in lattice parameter the austenitic matrix. This confers a uniquely low particle coarsening rate on the γ/γ' system, so that the alloy overages very slowly even at temperatures of the order of 0.7 T_m. Dislocation creep resistance is high because of the coherency strains set up between the precipitates and the matrix and because of the high temperature strength of the precipitate conferred by its ordered crystal structure. Solid solution strengthening by Cr, Co, Mo W and Ta is the other major source of high-temperature strength, and control of grain-boundary structure in super-alloys is largely practised through precipitation of substituted chromium carbides. Fig. 4.20 illustrates the relationship between stress–rupture properties and volume fraction of γ' for a variety of nickel-based super-alloys.

All γ'-hardened alloys show reduced strength at high temperatures because of the solution of the γ' precipitate, whereas alloys hardened by an insoluble dispersion of, say, an oxide do not have this characteristic. 'TD' nickel is an example of the latter type of alloy. Chemical reduction techniques are used for making the powders which are the feedstock of

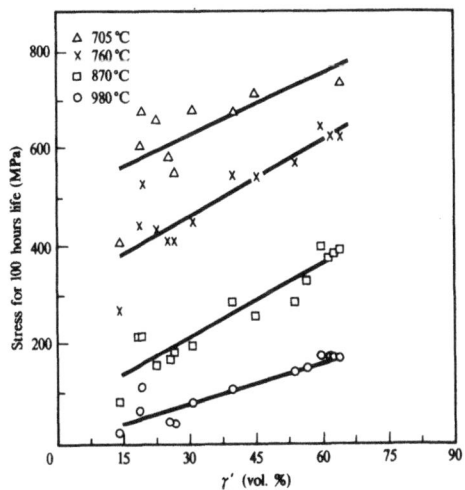

Fig. 4.20. The relationship between stress-rupture properties and volume fraction of γ-phase for a variety of nickel-based superalloys. (After Decker, 1969.)

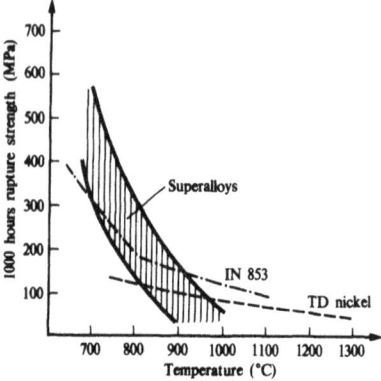

Fig. 4.21. Temperature dependence of the 1000 hour rupture strength of superalloys, TD nickel and the mechanically alloyed IN 853.

this material which typically contains about 2wt% of thoria, the particle diameter varying between 10 and 100 nm with interparticle spacing of less than 30 nm. A comparison is given in fig. 4.21 of the creep-rupture properties of TD nickel and typical wrought γ'-hardened superalloys. The sharp fall in strength with increase in temperature in the latter materials is obvious. TD nickel exhibits poor properties at intermediate temperatures, but there is only a modest decline in properties as the temperature rises (in contrast to the behaviour of the Nimonic superalloys), so that at temperatures above approximately 1100 K its properties are superior to those of the precipitation-hardened alloys. The development of the optimum properties in oxide-dispersion strengthened material depends critically upon the working and heat-treatment schedule used in their fabrication: the final microstructure is a very complex one consisting of finely-dispersed oxides in a matrix of fibrous grains containing a high density of dislocations in the form of a substructure.

It has more recently proved possible to produce a creep-resistant alloy which is strengthened not only by the presence of finely-divided particles of a stable oxide, but also by the γ' phase. They are prepared essentially by a powder route. The required composition is first prepared by blending in the appropriate materials - typically 20wt% Cr, 2.5wt% Ti, 1wt% Al and about 1wt% of a stable oxide such as alumina or yttria. The dry powders are then subjected to a high-energy ballmilling process using large nickel pellets. They are given a prolonged tumbling process which causes a continuous welding and fragmentation of the powders until they are practically homogeneous. Thus, mechanical energy rather than thermal energy promotes the intimate mixing, and the process is known as 'Mechanical alloying' (see, for example, Gessinger & Bomford, 1974).

Fig. 4.21 gives the creep-rupture data of such an alloy, and it is striking that good properties are obtained both at intermediate and at high temperatures. This powerful combination offers great opportunities for the designer of gas turbines.

It is obvious from the above discussion that by sophisticated alloy design, dislocation creep may be very effectively inhibited. Ashby (1973) has constructed a deformation map for a complex superalloy known as MAR-M200 of grain size 100 μm (fig. 4.22) which illustrates the effectiveness of the combination of many methods of strengthening. The alloy contains about 60% of Ni. The addition of W and Co provides solid-solution strengthening, and $Ni_3(Ti, Al)$ precipitation-strengthens. The alloy is used in the as-cast state (it is not capable of being rolled or forged) for blades in commercial gas turbines.

By comparison with fig. 4.19, the deformation map for pure Ni of the same grain size, it is clear that solution strengthening and precipitation hardening have raised the yield stress and have drastically curtailed the rate of power-law creep. The typical stresses and temperature to which a turbine blade is subjected is shown in figs. 4.19 and 4.22. If made of pure nickel, the blade would deform by power-law creep at an

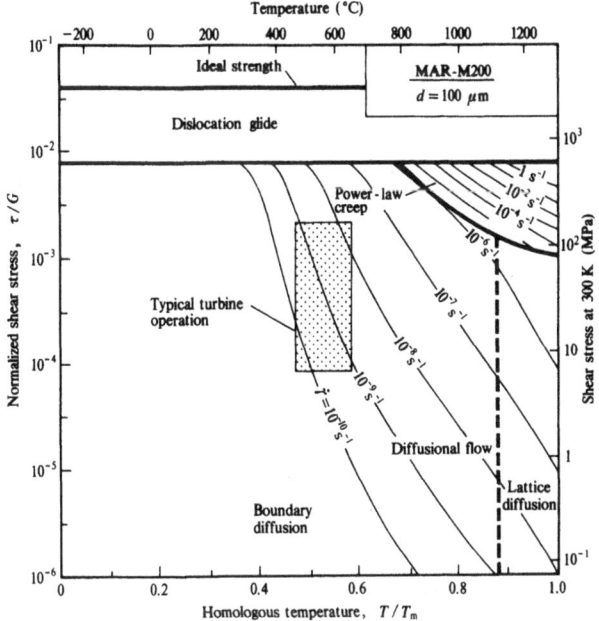

Fig. 4.22. The deformation mechanism map for the superalloy MAR-M200 of grain size 100 μm. The shaded box indicates the typical stresses and temperatures to which a turbine blade is subjected. (After Ashby, 1973.)

unacceptably high rate; in MAR-M200 the creep of the blade has been reduced by five orders of magnitude and the mechanism of flow has changed from power-law to diffusional creep.

4.4.2   Diffusional Creep in particle-hardened alloys

Fig. 4.22 suggests that in general, as dislocation creep is suppressed by sophisticated alloy design, diffusional creep will tend to dominate as components are exposed to higher and higher temperatures in service. It is well established that the deformation of pure metals at low stresses and elevated temperatures can occur by the stress-directed diffusion of vacancies. Lattice or grain-boundary diffusion can predominate depending on grain size and test temperature, and as indicated in the deformation mechanisms maps shown earlier, the corresponding creep rates $\dot{\epsilon}$ and $\dot{\epsilon}_B$ have been calculated to be

$$\dot{\epsilon} = Bp\Omega D/d^2 kT, \tag{4.20}$$

and

$$\dot{\epsilon}_B = B'p\Omega w D_B/d^3 \, kT, \tag{4.21}$$

where kT has the usual meaning, and

  p is the applied stress,
  Ω is the atomic volume,
  d is the grain size,
  w is the grain-boundary width,
  D is the lattice diffusion coefficient,
  D_B is the grain-boundary diffusion coefficient,
  B,B' are numerical constants which take the values 10 and $150/\pi$ respectively.

Equations (4.20) and (4.21) predict that the creep rates of materials of similar grain size vary only as the diffusion coefficient. However, recent experiments have shown that diffusional creep rates can be significantly inhibited by the presence of certain second-phase particles. Fig. 4.23 is a schematic representation of the distribution of precipitates in a cubic grain, ABCD, before and after diffusional creep deformation: the stress axis is vertical and the arrows inside the grain indicate the general direction of vacancy flow.

During deformation, high-angle grain boundaries that are normal to the stress axis emit vacancies, and boundaries that are parallel to this axis absorb vacancies. The number of precipitates intersected by unit area of the normal grain boundaries remains almost constant as the material is strained, whereas the parallel grain boundaries collect precipitates as

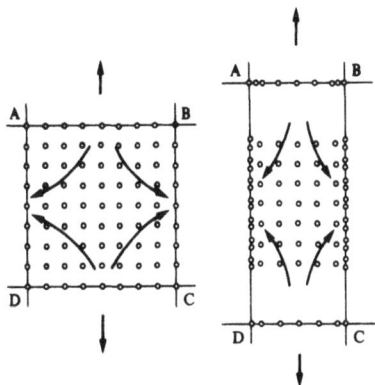

Fig. 4.23. Schematic representation of the effect of diffusional creep on particle distribution in a two-phase alloy.

vacancies are absorbed and the boundaries collapse, permitting the grains on either side to move towards each other as indicated in fig. 4.23. The normal boundaries thus acquire adjacent zones that are denuded of precipitates, while the parallel boundaries collect precipitates that would otherwise have been dispersed in a volume equivalent to that of the denuded zones. This process has been observed experimentally in, for example, specimens of a two-phase magnesium alloy 'Magnox ZR55' (Mg–0.55wt% Zr) by Squires, Weiner & Phillips (1963). Zones denuded of zirconium hydride precipitates were found adjacent to those grain boundaries which were normal to the applied tensile stress.

The inhibition of diffusional creep by the presence of a second phase has been clearly demonstrated by Harris *et al.* (1969), who creep-tested sintered Mg powder which contained dispersed particles of MgO. Their data, together with those for Magnox ZR55 for a test temperature of 400 °C is shown in fig. 4.24. The logarithm of the grain-size-compensated creep rate against the logarithm of the stress is plotted, and the theoretical creep rate according to (4.20) is shown (for a value of $B = 10$). Although it is clear that there is significant inhibition of diffusional creep in the ZR55 at low stresses, the creep rates of the sintered magnesium powder are as much as seven orders of magnitude lower than those predicted by the equation.

Since the presence of particles should not affect diffusion rates, their influence has been interpreted in terms of some limitation of grain boundaries as vacancy sources and sinks. There is also evidence that the inhibiting effect of the particles increases with increasing strain, and it has been proposed that this is due to the vacancy-absorbing capabilities of the parallel boundaries (fig. 4.23) being impaired by the progressive collection of precipitates. Two suggestions have been made (Harris *et al.*, 1969):

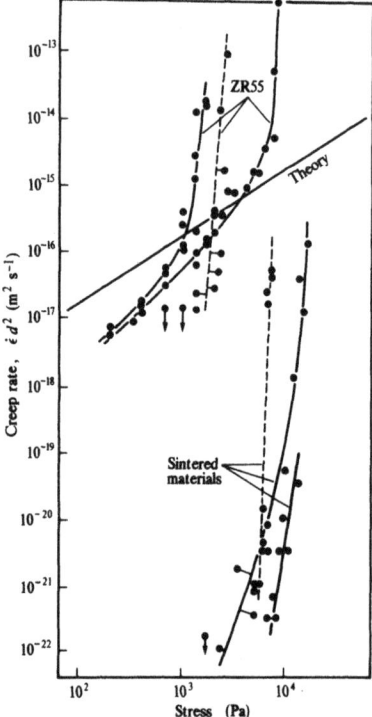

Fig. 4.24. Grain-size compensated creep rate plotted against stress for ZR55 and sintered magnesium products at 400 °C.

either the precipitates inhibit grain-boundary sliding and migration (which is a necessary component of diffusional creep deformation); or the precipitates restrict the ability of the parallel grain boundaries to collapse. This latter proposal has been considered in recent years by Harris (1973) and Burton (1973), and the model proposed to account for the effect by Harris is illustrated in fig. 4.25.

During diffusional creep, the annihilation of vacancies at the parallel grain boundaries will occur continuously. It is, however, instructive to imagine it as a discontinuous process, and in fig. 4.25 a parallel grain boundary containing precipitates has been allowed to *separate* as a result of the condensation of vacancies. Assuming that vacancy condensation cannot occur at the precipitate/matrix interfaces, then the grains on either side of the boundary cannot move towards each other unless the matrix deforms plastically around the precipitates. In fig. 4.25 such strain is achieved by the 'punching' of prismatic interstitial dislocation loops into the matrix. The loops are shown as decreasing in size with increasing distance from the precipitates owing to the fact that the loops will shrink

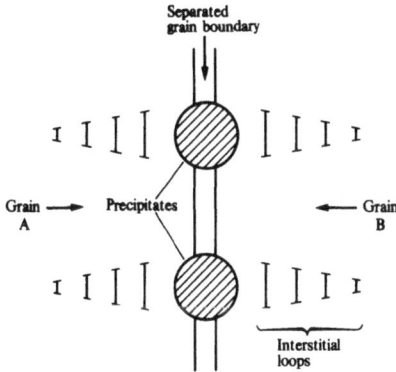

Fig. 4.25. The grain boundary between grains A and B has been allowed to separate as a result of the condensation of vacancies. In order that the two grains may move towards each other, thereby closing the gap, interstitial loops are generated in the matrix adjacent to the precipitates. (After Harris, 1973.)

by absorption of vacancies (thus compensating for the fact that condensation on the interface is not possible. The model predicts a 'threshold stress' for diffusional creep, which corresponds to the stress required to nucleate these dislocation loops. If it is assumed that creep deformation can only proceed by the generation of such dislocations in the matrix adjacent to the intergranular particles, a relationship may be derived between strain (ϵ) and time (t)

$$\epsilon = (\alpha/\beta) \, [1 - \exp(-\beta t)], \qquad (4.22)$$

where α and β are constants for a given specimen and test condition. The diffusional strain predicted from this equation was found to be in reasonable agreement with that observed in a creep specimen of ZR55.

The starting-point for the above hypothesis is that the diffusional properties of a two-phase interface are markedly different from those of a normal high-angle grain boundary. One suggestion is that at a two-phase interface, vacancies cannot be created or destroyed by the movement of atoms of *one* phase alone, but require the co-operative movement of the atoms comprising both phases. Thus, if a high melting-point phase exists in the grain boundary of a metal of low melting temperature, its interface with the matrix cannot act as an efficient source and sink for vacancies even at temperatures which approach the melting point of the matrix. Burton (1973) has summarized the reported data of two-phase systems, and fig. 4.26 shows his summary in the form of a volume fraction/relative melting point diagram, where the subscripts p and m refer to particle and matrix respectively. The regime where diffusional creep is

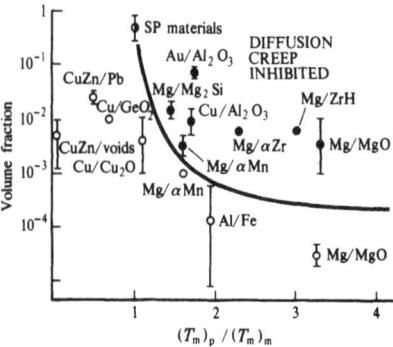

Fig. 4.26. Volume fraction/relative melting point diagram showing the regime where diffusion creep is inhibited. (After Burton, 1973.)

inhibited is indicated on the diagram, and the vertical bars on the points represent the range of compositions studied. It is clearly indicated in this figure that the conditions for inhibition are large volume fractions of elastically hard (high T_m) particles.

In conclusion, it must be stated that no direct experimental evidence has yet been obtained for the Harris (1973) and Burton (1973) model illustrated in fig. 4.25. In a recent paper, Nilsson, Howell & Dunlop (1979) favour the alternative explanation of diffusional creep inhibition by particles through their pinning action on grain-boundary sliding and migration. If it is accepted that grain-boundary sliding occurs by the movement of grain-boundary dislocations, then the latter will have to bypass the precipitates by Orowan looping or glide-climb around the particle/matrix interface. Hence the observed threshold stress may well be that stress which is necessary to move grain-boundary dislocations through the dispersion of intergranular precipitates.

4.4.3   Dislocation creep in particle-hardened alloys

Ansell & Weertman (1959) were the first to develop a quantitative model to describe the creep behaviour of dispersion-hardened alloys. The starting point of their theory is the assumption that the rate-controlling process is the climb of dislocations over the second-phase particles. The behaviour envisaged at *low* stress levels is that insignificant bowing or piling-up of dislocations occurs (i.e. the stress is less than the Orowan stress), and the derived expression for the creep rate $\dot{\epsilon}$ is

$$\dot{\epsilon} = \pi \, p \, b^3 \, D_s / 2kTd^2, \tag{4.23}$$

where kT has its usual meaning, and

D_s is the coefficient of self-diffusion of the matrix,
d is the particle diameter,

p   is the applied stress,

b   is the Burgers vector.

This expression thus predicts Newtonian behaviour, i.e. is proportional to p.

At stresses in excess of the Orowan stress, the dislocations are assumed to bow out and pinch off as loops around the particles. This occurs until the back-stress exerted by the loops around the particles prevents new dislocations from bowing to positions in between the particles. The rate of creep is then governed by the rate at which the innermost loop climbs off the particle and annihilates itself, and the concentric remaining loops collapse on to the particle, allowing a further dislocation to bow between the particles in order to pinch off and replace the lost loop. Rigorous analysis gives the expression for the creep rate as

$$\dot{\epsilon} = 2\pi p^4\ \lambda^2\ D_s/d\ G^3\ kT, \tag{4.24}$$

where G is the shear modulus and λ the interparticle spacing. Therefore, at high stresses the strain-rate is predicted to depend on the fourth power of the stress, and the activation energy is that appropriate to self-diffusion in the matrix.

The type of stress dependence predicted by (4.23) and (4.24) is not in agreement with experiments: in general the stress exponent is much higher than 4 for dispersion-hardened alloys. For example, in the case of thoriated nickel – 20wt% Cr single crystals tested in the temperature range 650 to 1300 °C (Lund & Nix, 1976), exponents ranging from 9 to 75 were observed. Experimental results for dispersion- and precipitation-hardened alloys indicate that the stress dependence of the creep rate may vary very widely between different materials and structural conditions. The theories of Ansell & Weertman are unable to account for this behaviour, and more recent models have been put forward which are based upon the process of recovery creep originally proposed by McLean (1966) for pure metals and solid solutions.

The model assumes that the dislocations formed during creep exist in a three-dimensional network. During primary creep the dislocation density increases. As the mesh-size in the dislocation network decreases with increasing dislocation density, the recovery rate accelerates and eventually a stage is reached where there is a balance between strain-hardening and recovery. This approach can lead to an expression for the steady-state creep rate of the form

$$\dot{\epsilon} = A\ p^n\ \exp\ (-Q/kT), \tag{4.25}$$

where n is the stress exponent, Q an activation energy, and A a structure-dependent quantity. For single-phase materials the value of n is usually close to 4, in agreement with theoretical predictions (Lagneborg, 1972). In an attempt to explain the high value of n found in particle-hardened materials, the particles and dislocation substructure have been considered to exert a back- or internal-stress (p_i) upon gliding dislocations and thereby reduce the creep-rate. Phenomenologically the creep rate due to dislocation motion is given by

$$\dot{\epsilon} = \phi \rho_m b \bar{v} \qquad (4.26)$$

where ρ_m is the density of mobile dislocations, b the Burgers vector, $\bar{v}$ the average dislocation velocity and ϕ an orientation factor. The dislocation velocity has generally been expressed in terms of an *average* internal stress, thus

$$\bar{v} = A \exp \{-[Q - V^* (p - p_i)/kT]\}, \qquad (4.27)$$

where Q is the activation energy for creep, V^* the activation volume and A a pre-exponential term. The mobile dislocation density is generally assumed to be a weaker function of the applied stress

$$\rho_m = \beta p^m, \qquad (4.28)$$

where β is a constant and m is between 2 and 3. In this approach the majority of the dislocations present are considered to be moving, and average values of ρ_m and $\bar{v}$ are considered.

This use of an average internal stress has been criticized (Gasca-Neri & Nix, 1974; Lloyd & McElroy 1974). The dislocation content of materials cannot be represented as a simple average density: it is necessary to consider the *distribution* of individual dislocation segment lengths, and this concept is illustrated in fig. 4.27. Here, the distribution of dislocation segment lengths in the network is described by a frequency function $N(l)$. The definition of $N(l)$ is such that there will be a number $N(l)dl$ of links in the range l and $(l + dl)$ within a unit volume of the material. In fig. 4.27 $N(l)$ is plotted as a function of the individual segment lengths (l): all dislocations in the material could at any moment be regarded as existing as links in the network, and the total dislocation density can be expressed as

$$\rho = \int_0^\infty l N(l) \, dl. \qquad (4.28)$$

On the application of a stress, p, those segments with flow stresses below p are capable of mobilization. The critical length l_c will be given by the equation $p = \alpha bG/l_c$ (where α is a constant). In fig. 4.27 l_{av} is also shown

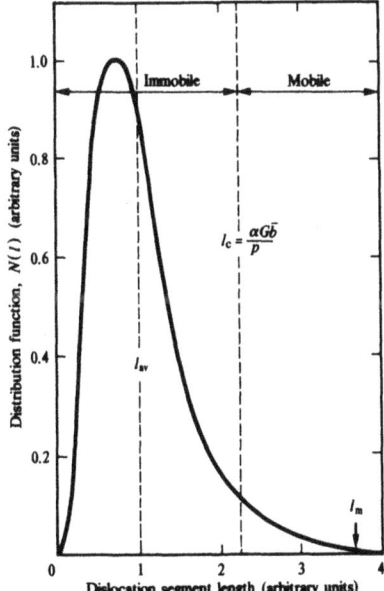

Fig. 4.27. The distribution $N(l)$ of dislocation segment lengths as a function of their individual lengths (l). The critical length l_c, given by $p = \alpha bG/l_c$, defines the mobile and immobile fractions; l_m is the maximum segment length and l_{av} the average segment length. (After Lloyd & McElroy, 1975.)

which is related to the average internal stress.

The number of segments which can be mobilized at any instant from a steady-state distribution is given by

$$\rho_m = \int_{l_c}^{\infty} f N(l) \, dl. \tag{4.30}$$

The majority of dislocations may thus be immobile at any given time, and the back-stress or internal stress produced by dislocation bowing has been used in recent models which take into account this distribution of dislocation link lengths. In this way Lagneborg (1972) has derived an expression for the steady-state creep rate as

$$\dot{\epsilon} = \int_{l_c}^{\infty} 4\pi b\lambda^2 \left[v_0 \frac{b}{l} \exp\left(- \frac{H_0 - bA\left[p - p_i(l_c)\right]}{kT} \right) \right] N(l) \, dl, \tag{4.31}$$

where λ is the mean free path of a dislocation segment, v_0 the Debye frequency, $p_i(l_c)$ is the internal stress caused by dislocations of length l_c, A

the activation area and H_0 the total activation energy. The expression is obtained by summing (over the entire range of link lengths) the product of the rate of link breakage and the incremental strain produced by subsequent link glide.

Because of the complexity of the problem, little systematic work has been performed on particle-hardened materials in creep with a view to correlation with the theoretical approach we have outlined. As discussed in §4.4.1, high-temperature alloys are very complex materials, for example, if manufactured by a powder-metallurgical route they will contain a high dislocation density, be polycrystalline, may contain both incoherent dispersoids and coherent γ'-type phases, as well as precipitates within the grain boundaries. Their evolution has moved in a largely empirical way and, for example, the relative merits of the presence of particles *per se*, compared with a stable dislocation substructure is a simple question not well understood at present.

One way of approaching the problem of attempting to identify the role of various features of such materials in affecting creep resistance, is to employ a model system which may be well characterized from a structural point of view. Thus, the use of single-crystal specimens will eliminate grain-boundary effects, and in such specimens stable dispersions of known characteristics may be introduced (e.g. by internal oxidation, as discussed earlier). Furthermore, by suitable mechanical–thermal processing, dislocation substructures may be introduced as well, so that the effect upon creep behaviour of the particle dispersion and of an additional dislocation substructure may be identified.

Creep in well-annealed particle-hardened crystals

Lloyd *et al.* (1973) examined the behaviour in creep of single crystals of copper containing dispersions of Al_2O_3 produced by internal oxidation, annealed to produce an initial low dislocation density (4×10^{10} m^{-2}). Fig. 4.28 shows the behaviour observed during the incremental loading of a crystal. No creep was detected at stresses below a critical value, which was always found to correlate with the yield stress p_Y determined at low strain rates. Incremental loading below p_Y produced only elastic strain increments, as did complete or partial unloading.

Exceeding the yield stress by a small amount gave rise to a net plastic extension as well as an elastic increment and, subsequently, to primary and secondary creep. One remarkable effect was that partial unloading to below p_Y resulted in continued creep at a reduced rate following elastic contraction and strain transient. The dislocation density after creep was 4.5×10^{13} m^{-2} (i.e. three orders of magnitude higher than before testing).

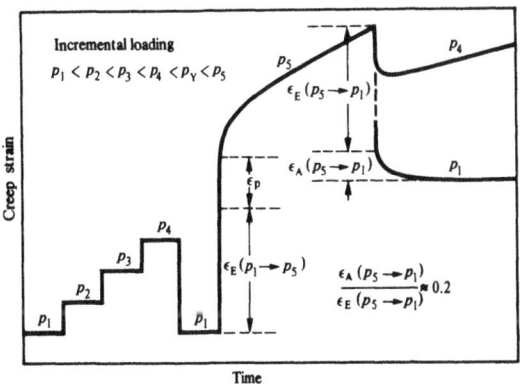

Fig. 4.28. Example of the incremental loading behaviour in creep of annealed upper crystal containing Al_2O_3 dispersion. (Courtesy of G. L. Lloyd.)

The stress dependence of the minimum creep rate may be illustrated in terms of the empirical relationship

$$\dot{\epsilon} = K' \exp(Bp) \exp(-Q'/kT), \tag{4.32}$$

where K' and B are materials constants, and Q' is a stress-independent activation energy. Fig. 4.29 shows data obtained from creep tests at 450 °C upon internally oxidized Cu-0.12wt% Al single crystals, plotted according to (4.32). It is seen that, in the specimen tested by reducing the creep stress to below p_Y after it had crept at a higher stress, the point corresponding to the observed creep rate falls on the same straight line as those obtained when testing above p_Y.

This effect can be explained in terms of the free dislocation lengths in the crystals as illustrated in fig. 4.27, and considering how changes in the mobile dislocation density will change the creep rate. In these well-annealed crystals the vast majority of dislocations exist as link lengths *between the particles*. The link length distribution is thus likely to be related to the scale of the particle-size distribution, provided the dislocation density is low enough, i.e. $\rho < \lambda^{-2}$, where λ is the mean particle spacing. Under these circumstances the variation of free dislocation segment lengths will be controlled solely by the particles. Increasing the dislocation density in this regime will cause a proportionate increase in creep rate, and this is the explanation of the appearance of creep strains in the specimen stressed below p_Y after its dislocation density had been increased by prior creep.

A similar effect has been reported in an Al-Cu-Mg alloy crept at 150 °C. Fig. 4.30 shows data due to Wilson (1973) at several stress levels for specimens with and without 3% prestrain (which introduced a high

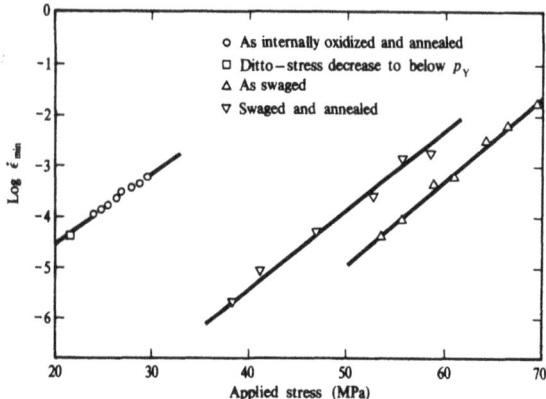

4.29. Creep behaviour of internally oxidised Cu–0.12 wt% Al single crystals at 450 °C. Data plotted according to (4.32). (Courtesy of G. J. Lloyd.)

density of dislocations that remained after the ageing treatment). In this system λ is very small, and hence the critical dislocation density (λ^{-2}) is very large.

The effect of high dislocation densities

Once the dislocation density exceeds λ^{-2}, dislocation nodes influence the segment length, so *decreasing* the number of mobile dislocations. The effect of cell and subgrain boundaries would be to reduce still further the number of mobile dislocations: this effect is known as 'substructure-strengthening'.

In view of the importance of dislocation density upon the creep behaviour in the copper single crystals described above, further tests were carried out on crystals containing deliberately introduced dislocation

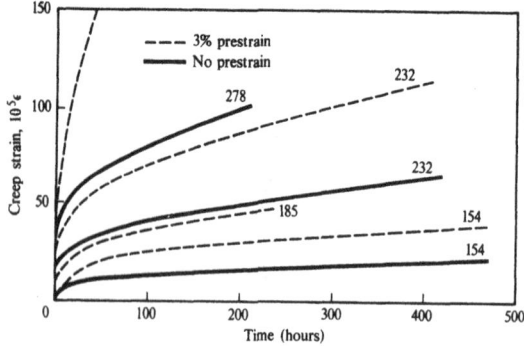

Fig. 4.30. The effect of 3% prestrain upon the creep curves of RR58 alloy at 150 °C. Numbers refer to the stress level in MPa. (After Wilson, 1973.)

substructures by swageing the crystals prior to testing. Swageing to 20% reduction in area produced a ragged cellular structure consisting of cell walls anchored at particles and a dislocation density in excess of 5×10^{14} m^{-2}, the cell-size being comparable with the mean particle spacing. After annealing, a sharper subgrain structure was present, the dislocation density having been reduced to 10^{-14} m^{-2} without any sign of recrystallization.

In contrast to the well-annealed crystals, these specimens exhibited no creep limit (as in fig. 4.28). The data are shown in fig. 4.29. In terms of (4.32) it is clear that B is approximately constant for the material under the various conditions, but the K' exp $(-Q'/kT)$ term varies widely with treatment. Fig. 4.29 demonstrates the dramatic effect of dislocation substructure strengthening upon creep resistance in comparison with particle-strengthening. In the swaged specimens the effect of the particles upon conferring creep strength is thus indirect, in that they inhibit the recovery of the dislocation substructure.

The exponential stress dependence shown in fig. 4.29 may be accounted for in terms of (4.26) and (4.30), since the form of $N(l)$ is likely to be Gaussian (1.37). This would lead to an exponential increase in ρ_m with applied stress via (4.30).

In conclusion it may be stated that, in order to maximize the creep deformation resistance of a precipitation hardening alloy, three structural criteria should be met. Firstly, there should be a fine three-dimensional dislocation network in order to maximize the flow stress. Secondly, the recovery rate should be minimized by ensuring that a large fraction of the precipitate particles lie on dislocations in order to anchor them. Coarsening of the dislocation network is expected to start, as with grain growth, at local instabilities. The third requirement is therefore that the dislocation network should be as *uniform* as possible.

McElroy *et al.* (1974) attempted to maximize the creep resistance of a precipitation-hardening fcc iron alloy, and concluded that a multiple mechanical thermal treatment (mmtt) seemed most likely to satisfy the three criteria mentioned above. The treatment consisted of repeated cycles, each cycle comprising a small strain at room temperature and an ageing treatment at a higher temperature. Tests showed that mmtt samples have substantially greater creep deformation resistance than samples treated in other ways, although a sharp drop in ductility after mmtt was reported, perhaps because the particular alloy used was of low inherent cohesion.

4.5   High-temperature fracture in particle-hardened alloys

We have seen that the creep strength in structural materials for high-temperature applications is usually obtained by the dispersion of fine second-phase particles which inhibit dislocation mobility, and by the precipitation of particles in the grain boundary which obstruct sliding and may inhibit diffusion-creep processes. It is commonly observed, however, that an increase in creep strength is accompanied by a decrease in the fracture toughness of the material at elevated temperature, and it has been suggested that a situation has almost been reached where the fracture toughness rather than creep resistance has become the limiting factor in the service life of these materials.

As with fracture at low temperature, creep fracture could involve either *rupture, transgranular fracture* or *intergranular fracture*. As discussed in chapter 3, ductile rupture as illustrated in figs. 3.6*a* and *b* does not normally proceed in particle-hardened alloys up to strains corresponding to 100% reduction in area, since cavities nucleate and grow at the particles leading to fracture by micro-void coalescence at considerably smaller strains.

4.5.1 Transgranular rupture

Rupture by 100% reduction in area is less uncommon after high-temperature deformation, and this is because the nucleation of internal voids is suppressed. This can happen in a number of ways.

(i) With increase in temperature the second-phase particles may go into solid solution. A sudden onset of rupture would be observed on crossing the solvus line in this circumstance.

(ii) High rates of local recovery will relieve the stress at inclusions, preventing it from reaching the level required for nucleation, as discussed in §3.3.1. Dynamic recrystallization appears to permit abnormally rapid local stress-relief, and can induce rupture in many fcc metals and alloys, and this is reported in a paper by Pavinich & Raj (1977) who studied the fracture modes at elevated temperatures of polycrystalline copper containing a dispersion of silica particles, the principal variables being grain size, strain-rate, and the size and spacing of the second-phase particles. Pavinich & Raj observed that at high strain-rates their samples fractured by necking to a point without any cavitation, and there was found to be extensive recrystallization within the necked region.

4.5.2 Intergranular fracture

Pavinich & Raj observed three modes of failure in their material. In addition to the transgranular necking fracture referred to above, they also found fracture by the propagation of intergranular cracks initiated at the surface, and also fracture by grain-boundary cavitation throughout the entire specimen cross-section. The transition between the fracture modes was shown to shift systematically with temperature, strain-rate and the microstructural parameter d/λ, where d is the average particle size and λ the average spacing in the grain boundaries. This parameter constitutes an area fraction. Fig. 4.31 illustrates the variation of the critical strain-rate at which the transitions in fracture mode occurred at different temperatures as a function of d/λ. At a constant temperature, the critical strain-rate increases with increasing area fraction of the second phase, and also, for a given area fraction of the phase, the critical strain-rate increases with increasing temperature.

These results show that a higher temperature and a larger second-phase content favour intergranular fracture, although no definite explanation for this behaviour is offered, since further experiments appear to be needed in this area. Pavinich & Raj's results for intergranular fracture of copper containing SiO_2 particles by the nucleation and growth of intergranular voids are summarized in fig. 4.32, which shows the observed variation in ductility of their material with d/λ at different temperatures. Initially ductility at constant temperature decreased drastically with increasing d/λ, while further increases in d/λ produced only small decreases in ductility. In this context these authors define 'ductility' in terms of the rate equations for the time-to-fracture (t_f) and the steady-state creep strain-rate ($\dot{e}$), which had similar dependence on stress:

$$t_f = A' p^{-n},$$
$$\dot{e} = A p^n,$$

(4.33)

where the constants A and A' depended upon temperature and the microstructure. The ductility was defined as $\dot{e} t_f$.

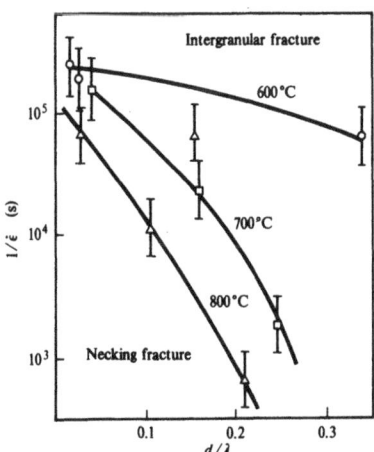

Fig. 4.31. Variation of the critical strain-rate at which fracture occurs by crack propagation with the parameter d/λ at different temperatures. (After Pavinich & Raj, 1977.)

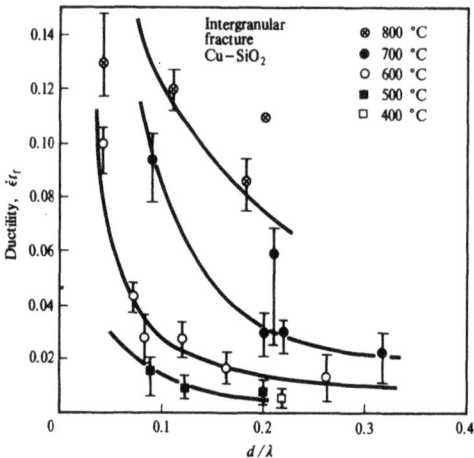

Fig. 4.32. Variation of ductility, $\dot{\epsilon}t_f$, with d/λ at different temperatures for Cu-SiO$_2$. (After Pavinich & Raj, 1977.)

A model is proposed to explain the form of the curves in fig. 4.32, to establish a lower bound and a practical upper bound for ductility when fracture occurs by the coalescence of a row of cavities in the grain boundary. Pavinich & Raj consider a two-dimensional polycrystal of grain size D, containing holes of size h spaced a distance l from each other (fig. 4.33). The area fraction of holes in the grain boundary will be h/l, their initial values being h_0/l_0. One may consider deformation in the hole region αl, where the deformation will be inhomogeneous, and in the homogeneous region $(D - \alpha l)$. The term α would be expected to vary with strain-rate sensitivity, $1/n$. On physical grounds α will increase with decreasing n, so that the maximum value of α will correspond to minimum value of n which is unity. When $n = 1$, this corresponds to a Newtonian viscous fluid, and it is calculated in this case that $\alpha = 3$. The lower bound value of α, which corresponds to $n \to \infty$ is assumed to be $\alpha_{min} = 1$ (see fig. 3.8b).

If the strain is ϵ' in the hole region and is ϵ'' in the homogeneous region, then the total strain is given by

$$\epsilon = \frac{\alpha l}{D}(\epsilon' - \epsilon'') + \epsilon''. \tag{4.34}$$

To calculate the lower bound of ductility, ϵ_{lb}, it is assumed that the deformation is completely localized, i.e. $\epsilon'' = 0$, and $\alpha = 1$, then

$$\epsilon_{lb} = \frac{l}{D}\epsilon'. \tag{4.35}$$

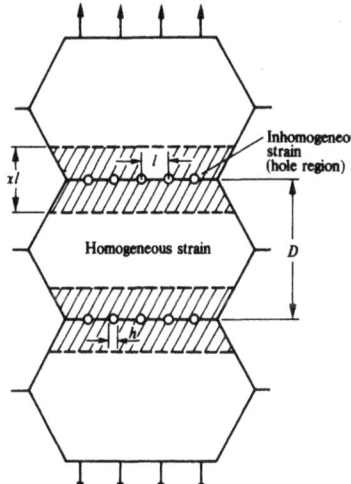

Fig. 4.33. Size and spacing of holes in a polycrystal of grain size D.

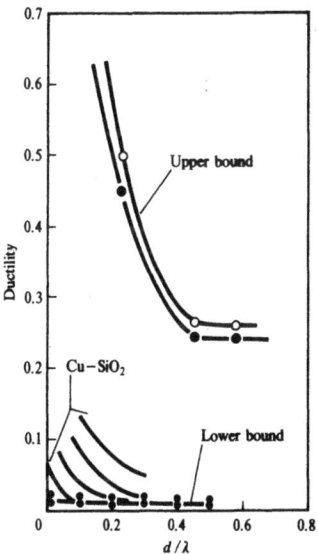

Fig. 4.34. The upper bound and lower bound of ductility for different ratios of d/λ. (After Pavinich & Raj, 1977.)

This ductility was calculated as a function of d/λ by assuming that the initial porosity $h_0/l_0 \approx d/\lambda$ and by estimating the ϵ' required to coalesce the cavities, and by substituting that result in (4.35). Plots of ϵ_{lb} for λ/D

$= \frac{1}{20}$ and $\frac{1}{40}$ are shown in fig. 4.34 (these values of λ/D correspond to the ratio of particle spacing to grain size in their $Cu-SiO_2$ specimens).

The upper bound of ductility was then calculated by using $\alpha = 3$, and measured values ϵ' and ϵ'' obtained experimentally in a model system. Substitution in (4.34) gave the upper bound curves shown in fig. 4.34.

The calculated curves thus have a similar form to the experimental curves, and the results are also in qualitative agreement with the data on nickel-based alloys and other materials published in the literature. Thus it appears that it is possible to treat the growth of cavities in a grain boundary at elevated temperature as a power-law creep type of matrix plasticity. However, analytical solutions for the growth of a row of cavities in a plastically deforming continuum are not at present available for a general value of the power-law stress exponent, n.

Finally, it should be pointed out that when the temperature is sufficiently high, holes on grain boundaries in stressed solids can grow by *diffusion*, and the diffusional growth should be regarded as an alternative to the power-law creep discussed above. There appears to be little experimental evidence of diffusional growth being a dominant mechanism, although the problem has been considered theoretically (Hull & Rimmer, 1959; Raj & Ashby 1975).

References

Chapter 1

Aaron, H.B., Fainstein, D. & Kotler, G.R. (1970) *J. Appl. Phys.*, **41**, 4404.

Ashby, M.F. & Ebeling, R. (1966) *Trans. AIME*, **236**, 1396.

Cahn, J.W. (1956) *Acta Met.*, **4**, 449.

Cahn, J.W. (1957) *Acta Met.*, **5**, 168.

Cahn, J.W. & Hilliard, J.E. (1958) *J. Chem. Phys.*, **28**, 258.

Cahn, J.W. & Hilliard, J.E. (1959) *J. Chem. Phys.*, **31**, 688.

Christian, J.W. (1975) *The Theory of Transformations in Metals and Alloys*, Pt 1 (2nd Edition), Pergamon, Oxford.

Corti, C.W., Cotterill, P. & Fitzpatrick, G.A. (1974) *Int. Met. Rev.*, **19**, 77.

Exner, H.E., Santa Marta, E. & Petzow, G. (1971) *Modern Developments in Powder Metallurgy* (Ed. H. H. Hausner), Plenum Press, New York, Vol. 4, p. 315.

Fink, W.L. & Willey, L.A. (1947) *Metals Technology*, p. 2225.

Foreman, A.J.E. & Makin, M.J. (1966) *Phil. Mag.*, **14**, 911.

Fullman, R.L. (1953) *Trans. AIME*, **197**, 447.

Gibbs, J.W. (1878) *Trans. Conn. Acad.*, **3**, 102, 345.

Hornbogen, E. (1967) *Aluminium*, **43**, 115.

Hornbogen, E. (1969) *Nucleation* (Ed. A. C. Zettlemoyer), Dekker, New York, p. 309.

Hornbogen, E. & Roth, M. (1967) *Z.F. Metallk.*, **58**, 842.

Jacobs, M.H. & Pashley, D.W. (1969) *The Mechanism of Phase Transformation in Crystalline Solids*, Institute of Metals, London, Monograph no. 33, p. 43.

Kocks, U.F. (1966) *Phil. Mag.*, **14**, 1629.

Kinsman, K.R. & Aaronson, H.I. (1967) *Transformation and Hardenability in Steels*, Climax Molybdenum Co., Ann. Arbour, Michigan, USA, p. 39.

Kirchner, H.P.K. (1971) *Met. Trans.*, **2**, 2861.

Köster, U. (1971) *Recrystallization of metallic materials* (Ed. F. Haessner) Riederer, Stuttgart, p. 215.

Lifshitz, I.M. & Slyozov, V.V. (1961) *J. Phys. Chem. Solids*, **19**, 35.

Martin, J.W. & Doherty, R.D. (1976) *Stability of Microstructure in Metallic Systems*, Cambridge University Press.

Meyrick, G. (1976) *Scripta Met.*, **10**, 649.

Nabarro, F.R.N. (1940) *Proc. Roy. Soc.*, **175**, 519.

Polmear, I.J. (1968) *J. Inst. Metals*, **20**, 20.

Russell, K.C. & Aaronson, H.I. (1975) *J. Mat. Sci.*, **10**, 1991.

Servi, I.S. & Turnbull, D. (1966) *Acta Met.*, **14**, 161.

Shepherd, J.P. (1969) *Met. Sci. J.*, **3**, 229.

Stewart, A.T. & Martin, J.W. (1970) *J. Inst. Metals*, **98**, 62.

Stickler, R. & Vinckier, A. (1963) *Mem. Sci. Rev. Met.*, **60**, 2.

Tu, K. & Turnbull, D. (1967) *Acta Met.*, **15**, 369.

Wagner, C. (1961) *Z. Elektrochem.*, **65**, 581.

Chapter 2

Anand, L. & Gurland, J. (1975) *Met. Trans.*, 6A, 928.

Anand, L. & Gurland, J. (1976a) *Met. Trans.*, 7A, 191.

Anand, L. & Gurland, J. (1976b) *Acta Met.*, 24, 901.

Ashby, M.F. (1966) *Acta Met.*, 14, 679.

Ashby, M.F. (1971) *Strengthening Methods in Crystals* (Ed. A. Kelly & R.B. Nicholson), Elsevier, Amsterdam, p. 137.

Brown, L.M. & Ham, R.K. (1971) '*Strengthening Methods in Crystals*' (Ed. A. Kelly & R. B. Nicholson), Elsevier, Amsterdam, p. 12.

Brown, L.M. & Stobbs, W.M. (1971) *Phil. Mag.*, 23, 1185, 1201.

Chaturvedi, M.C., Lloyd, D.J. & Chung, D.W. (1976) *Met. Sci. J.*, 10, 373.

Dew-Hughes, D. & Robertson, W.D. (1960) *Acta Met.*, 8, 147.

Duwez, P. (1967) *Trans. Amer. Soc. Metals*, 60, 607.

Eshelby, J.D. (1957) *Proc. Roy. Soc.*, A241, 376.

Foreman, A.J.E. & Makin, M.J. (1967) *Canad. J. Phys.*, 45, 511.

Gould, D. (1971) D. Phil. thesis, Oxford University.

Harkness, S.D. & Hren, J.J. (1970) *Met. Trans.*, 1, 43.

Hazzledine, P.M. & Hirsch, P.B. (1974) *Phil. Mag.*, 30, 1331.

Hirsch, P.B. & Humphreys, F.J. (1970) *Proc. Roy. Soc.*, A318, 45.

Hirsch, P.B. & Kelly, A. (1965) *Phil. Mag.*, 12, 881.

Hornbogen, E. & Staniek, G. (1974) *J. Mat. Sci.*, 9, 879.

Humphreys, F.J. (1977) *Acta Met.*, 25, 1323.

Humphreys, F.J. & Stewart, A.T. (1972) *Surface Sci.*, 31, 389.

Jones, R.L. & Kelly, A. (1968) *Proceedings of the 2nd Bolton Landing Conference on Oxide Dispersion Strengthening* (Ed. G. S. Ansell), Gordon and Breach, New York.

Karlsson, B. & Lindén, G. (1975) *Mat. Sci. Eng.*, 17, 209.

Kelly, P.M. (1973) *Int. Met. Rev.*, 18, 31.

Melander, A. & Persson, P.A. (1978) *Acta Met.*, 26, 267.

Nicholson, R.B. (1971) '*Strengthening Methods in Crystals*' (Ed. A. Kelly & R. B. Nicholson), Elsevier, Amsterdam, p. 535.

Russell, K.C. & Ashby, M.F. (1970) *Acta Met.*, 18, 891.

Staniek, G. & Hornbogen, E. (1973) *Scripts Met.*, 7, 615.

Stewart, A.T. & Martin, J.W. (1975) *Acta Met.*, 23, 1.

Törrönen, K. (1976) *Proceedings of the Fourth International Conference on the Strength of Metals and Alloys* (Nancy), Metals Society, p. 239.

Witt, M. & Gerold, V. (1969) *Scripta Met.*, 3, 371.

Chapter 3

Albrecht, J., Martin, J.W.R., Lütjering, G. & Martin, J.W. (1976) *Proceedings of the Fourth International Conference on the Strength of Metals and Alloys* (Nancy), Metals Society, p. 463.

Argon, A.S., Im, J. & Safoglu, R. (1975) *Met. Trans.*, 6A, 825.

Ashby, M.F. (1966) *Phil. Mag.*, 14, 1157.

Ashby, M.F. (1977) Fracture Mechanism Maps *CUED/C/MATS/TR.* 34.

Averbach, B.L. (1974) *Fracture Prevention and Control: Proceedings of a Symposium at the 1972 Western Metal and Tool Exposition and Conference* (Los Angeles), (Ed. D. W. Hoeppner), American Society for Metals, Metals Park, Ohio, p. 97.

Benson, J.P. & Edmonds, D.V. (1977) *Fracture 1977: Proceedings of the Fourth*

International Conference on Fracture (Waterloo, Canada), (Ed. D.M.R. Taplin), Bonfield, W. (1972) *Scripta Met.*, 6, 77.

Brown, L.M. & Embury, J.F. (1973) *The Microstructure and Design of Alloys: Proceedings of the Third International Conference on the Strength of Metals and Alloys* (Cambridge), The Institute of Metals, Vol. 1, p. 164.

Calabrese, C. & Laird, C. (1974) *Mat. Sci. Eng.*, 13, 14.

Clayton, J.Q., & Knott, J.F. (1976) *Metal Science*, 10, 63.

Cottrell, A.H. (1958) *Trans. AIME*, 212, 192.

Cox, T.B. & Low, J.R. (1974) *Met. Trans.*, 5, 459.

Dowling, J.M. & Martin, J.W. (1973) *The Microstructure and Design of Alloys: Proceedings of the Third International Conference on the Strength of Metals and Alloys* (Cambridge), The Institute of Metals, Vol. 1, p. 170.

Edelson, B.I. & Baldwin, W.M. (1962) *Trans. Amer. Soc. Met.*, 55, 230.

Evans, A.G. (1972) *Phil. Mag.*, 26, 1327.

Fine, M.E. & Santner, J.S. (1975) *Scripta Met.*, 9, 1239.

Forsyth, P.J.E. (1963) *J. Australian Inst. Metals*, 8, 52.

Garrett, G.G. & Knott, J.F. (1976) *Proceedings of the Second International Conference on the Mechanical Behaviour of Materials* (Boston, Mass.).

Goods, S.H. & Brown, L.M. (1979) *Acta Met.*, 27, 1.

Gurland, J. & Plateau, J. (1963) *Trans. Amer. Soc. Met.*, 56, 442.

Hahn, G.T. & Rosenfield, A.R. (1968) *ASTM*, STP 432 p. 5.

Hahn, G.T. & Rosenfield, A.R. (1975) *Met. Trans.*, 6A 653.

Hornbogen, E. (1975) *Z.F. Metallk.*, 66, 511.

Hornbogen, E. & Lütjering, G. (1975) *Proceedings of the Sixth International Conference on Light Metals* (Leoben, Vienna), Al-Verlag, Düsseldorf.

Hornbogen, E. & Zum Gahr, K.H. (1975) *Metallography*, 8, 181.

Hornbogen, E & Zum Gahr, K.H. (1976) *Acta Met.*, 24, 581.

Kawabata, T. & Izumi, O. (1976) *Acta Met.*, 24, 817.

Knott, J.F. (1973) *Fundamentals of Fracture Mechanics*, Butterworths, London.

Knott, J.F. (1977) *Fracture 1977: Proceedings of the Fourth International Conference on Fracture* (Waterloo, Canada), (Ed. D. M. R. Taplin), University of Waterloo Press, Canada, Vol. 1, p. 61.

Kotilainen, H. & Törrönen, K. (1977) *Fracture 1977: Proceedings of the Fourth International Conference on Fracture* (Waterloo, Canada), (Ed. D. M. R. Taplin), University of Waterloo Press, Canada, Vol. 2, pp. 57, 141.

Laird, C. (1976) *Mat. Sci. Eng.*, 25, 187.

Lange, F.F. (1971) *J. Amer. Ceram. Soc.*, 49, 614.

Lawn, B.R. & Wilshaw, T.R. (1975) *Fracture of Brittle Solids*, Cambridge University Press.

Lindley, T.C., Richards, C.E. & Ritchie, R.O. (1975) *Proceedings of the Conference on the Mechanics and Physics of Fracture* (Cambridge), Institute of Physics, Metals Society.

Lu, M.C. & Weissmann, S. (1978) *Mat. Sci. Eng.*, 32, 41.

Lütjering, G. & Weissmann, S. (1970) *Acta Met.*, 18, 785.

McClintock, F.A. (1968) in *Ductility*, American Society for Metals, Metals Park, Ohio, p. 255.

McEvily, A.J., Clark, J.B., Utley, E.C. & Herrnstein, W.H. (1963) *Trans. AIME*, 227, 1093.

McGrath, J.T. & Bratina, T.W.J. (1967) *Acta Met.*, 15, 329.

McGrath, J.T. & Bratina, T.W.J. (1970) *Phil. Mag.*, 21, 1087.

McMahon, C.J., & Cohen, M. (1965) *Acta Met.*, 13, 591.

Neumann, P. (1974) *Acta Met.*, **22**, 1155, 1167.

Ostermann, F.G. (1971) *Met. Trans.*, **2**, 2897.

Paris, P.C. & Erdogan, F. (1963) *J. Basic Eng.* (*Trans. ASME*, D), **85**, 528.

Rawal, S.P. & Gurland, J. (1977) *Fracture 1977: Proceedings of the Fourth International Conference on Fracture* (Waterloo, Canada), (Ed. D. M. R. Taplin), University of Waterloo Press, Canada, Vol. 2, p. 41.

Rice, J.R. & Johnson, M.A. (1970) *Inelastic Behaviour of Solids*, (Ed. M.F. Kanninen *et al.*), McGraw-Hill, New York, p. 641.

Ritchie, R.O., Knott, J.F. & Rice, J.R. (1973) *J. Mech. Phys. Sol.*, **21**, 395.

Smith, E. (1966) *Physical Basis of Yield and Fracture: Conference Proceedings* (Oxford), Institute of Physics, Physical Society, p. 36.

Smith, G.C. (1975) *Proceedings of the Conference on the Mechanics and Physics of Fracture* (Cambridge), Institute of Physics, Metals Society.

Tanaka, K., Mori, T. & Nakamura, T. (1970) *Phil. Mag.*, **21**, 267.

Chapter 4

Ansell, G.S. & Weertmen, J. (1959) *Trans. AIME*, **215**, 838.

Ashby, M.F. (1973) *The Microstructure and Design of Alloys: Proceedings of the Third International Conference on the Strength of Metals and Alloys* (Cambridge), The Institute of Metals, Vol. 2, p. 8.

Brown, L.M. & Ham, R.K. (1971) *Strengthening Methods in Crystals*, (Ed. A. Kelly & R. B. Nicholson), Elsevier, Amsterdam, p. 9.

Burton, B. (1973) *Mat. Sci. Eng.*, **11**, 337.

Decker, R.F. (1969) in *Steel-Strengthening Mechanisms*, Climax Molybdenum Co., Zurich, p. 147.

Doherty, R.D. & Martin, J.W. (1962-3) *J. Inst. Metals*, **91**, 332.

Friedel, J. (1964) *Dislocations*, Pergamon Press, Oxford.

Gasca-Neri, R. & Nix, W.D. (1974) *Acta Met.*, **22**, 257.

Gawne, D.T. & Higgins, G.T. (1971) *J. Mat. Sci.*, **6**, 403.

Gessinger, G.H. & Bomford, M.J. (1974) *Int. Met. Rev.*, **19**, 51.

Gould, D., Hirsch, P.B. & Humphreys, F.J. (1974) *Phil. Mag.*, **30**, 1353.

Hansen, N. (1975) *Mem. Sci. Rev. Met.* (March) 189.

Hansen, N. & Bay, B. (1972) *J. Mat Sci.*, **7**, 1351.

Harris, J.E. (1973) *Met. Sci. J.*, **7**, 1.

Harris, J.E., Jones, R.B. Greenwood, G.W. & Ward, M.J. (1969) *J. Australian Inst. Metals*, **14**, 154.

Hillert, M. (1965) *Acta Met.*, **24**, 391.

Hirsch, P.B. & Humphreys, F.J. (1969) *Physics of Strength and Plasticity*, (Ed. A. S. Argon), MIT Press, Cambridge, Mass. p. 189.

Hull, D. & Rimmer, D.E. (1959) *Phil. Mag.*, **4**, 673.

Humphreys, F.J. (1977) *Acta Met.*, **25**, 1323.

Humphreys, F.J., Hirsch, P.B. & Gould, D. (1970) *Proceedings of the Second International Conference on the Strength of Metals and Alloys* (Asimolar), American Society for Metals, p. 550.

Kamma, C. & Hornbogen, E. (1976) *J. Mat. Sci.*, **11**, 2340.

Köster, U. (1974) *Mat. Sci. J.*, **8**, 151.

Lagneborg, R. (1972) *Int. Met. Rev.*, **17**, 130.

Lloyd, G.J. & McElroy, R.J. (1974) *Acta Met.*, **22**, 339.

Lloyd, G.J., McElroy, R.J. & Martin, J.W. (1973) *The Microstructure and Design of Alloys: Proceedings of the Third International Conference on the Strength of*

Metals and Alloys (Cambridge), The Institute of Metals, Vol. 1, p. 185.

Lund, R.W. & Nix, W.D. (1976) *Acta Met.*, 24, 469.

Mäder, K., & Hornbogen, E. (1974) *Scripta Met.*, 8, 979.

McElroy, R.J. Ishida, Y., McLean, D. & Szkopiak, Z. (1974) *Metals Technology*, 1, 468.

McLean, D. (1966) *Reports on Progress in Physics*, 29, 1.

Nes, E. (1976a) *Acta Met.*, 25, 1323.

Nes, E. (1976b) *Scripta Met.*, 10, 1025.

Nilsson, J.-O., Howell, P.R. & Dunlop, G.L. (1979) *Acta Met.*, 27, 179.

Pavinich, W. & Raj, R. (1977) *Met. Trans.*, 8A, 1917.

Raj, R. & Ashby, M.F. (1975) *Acta Met.*, 23, 653.

Rollason, T.C. & Martin, J.W. (1970) *Acta Met.*, 18, 1267.

Scharf, G. & Gruhl, W. (1969) *Z.f. Metallk.*, 60, 413.

Shewfelt, R.S.W. & Brown, L.M. (1977) *Phil. Mag.*, 35, 945.

Smith, C.S. (1948) *Trans. AIME*, 175, 345.

Squires, R.L., Weiner, R.T. & Phillips, M. (1963) *J. Nucl. Mat.*, 8, 77.

Stewart, A.T. & Martin, J.W. (1975) *Acta Met.*, 23, 1.

Vasudevan, A.K., Petrovic, J.J. & Roberson, J.A. (1974) *Scripta Met.*, 8, 861.

Weertman, J. & Weertman, J.R. (1965) *Physical Metallurgy*, (Ed. R.W. Cahn) North-Holland Publishing Co., Amsterdam, Ch. 16.

Wilson, R.N. (1973) *J. Inst. Metals*, 101, 188.

Index

For EU product safety concerns, contact us at Calle de José Abascal, 56–1°, 28003 Madrid, Spain or eugpsr@cambridge.org.

www.ingramcontent.com/pod-product-compliance
Ingram Content Group UK Ltd.
Pitfield, Milton Keynes, MK11 3LW, UK
UKHW022108141225
466003UK00014B/638